SpringerBriefs in Biochemistry
and Molecular Biology

More information about this series at http://www.springer.com/series/10196

Alka Dwevedi

Protein Folding

Examining the Challenges from Synthesis to Folded Form

 Springer

Alka Dwevedi
UNESCO-Regional Centre
  for Biotechnology
Gurgaon
India

ISSN 2211-9353   ISSN 2211-9361 (electronic)
SpringerBriefs in Biochemistry and Molecular Biology
ISBN 978-3-319-12591-6   ISBN 978-3-319-12592-3 (eBook)
DOI 10.1007/978-3-319-12592-3

Library of Congress Control Number: 2014957947

Springer Cham Heidelberg New York Dordrecht London

Printed on acid-free paper

Springer is part of Springer Science+Business Media (www.springer.com)

Preface

The primary structure of a protein containing sequence of amino acids has the blue-print for the complete stable and functional tertiary structure. However, there is a huge gap in our knowledge of how we move from protein sequence to function in living organisms. Lots of efforts are being made to predict three-dimensional structure of the protein from its primary structure. It is strongly believed that deep understanding of protein folding would be the way to solve the problem. Further, it would also help to understand diverse cellular processes; *viz*. transfer of information by the ligand through specific receptor toward cell's interior, generation of antibodies against specific antigens, etc. as well as giving solution to the problem of protein misfolding leading to number of diseases. The present book focuses on major challenges in protein folding, approaches made, and it's till date research updates. This will help in improved understanding of the route that a protein takes from its synthesis to the correct folded form.

Chapter 1: The principle guiding protein to fold into its native conformation has been searched by biochemists, structural biologists, molecular biologists and recently by bioinformaticians. The problem is not new but it is being worked for approximately 100 years. There are various challenges, most importantly short-lived folding intermediates and formation of inter-residual interactions including disulfide linkages. Various experimental and theoretical models are being generated to decipher the pathways and principles of protein folding but none have the complete solution to the problem. The present chapter is the compilation of important research contributions in the arena of protein folding made till date.

Chapter 2: The protein is functionally active in its three-dimensional native state. Almost all cells' machinery is based on the involvement of number of proteins. Native state of the protein is the most active and stable state with a specific conformation determined by polypeptide backbone and intermolecular interactions. Efficient protein folding is the prerequisite of proper functioning of cell machinery which depends on coordinated functioning of chaperones, chaperonins, and various auxiliary cofactors. The failure of a specific protein to adopt its native and active state is called "protein misfolding". Protein misfolding has wide range of pathological implications due to loss of normal cellular functions. The present chapter is based on the criteria determining proper protein folding based on their structural and kinetics studies.

Chapter 3: Protein folding has now been accepted to be a self assembly process with favorable environmental parameters. Native state of the protein has lowest free energy due to various interactions (ionic, covalent, and hydrogen bonds and hydrophobic interactions). From last decades, researchers are trying to predict *de novo* three-dimensional structure of a protein from its amino acid sequence using parameters of folding energetics. Further, efforts are also being made to understand protein–protein and protein–ligand interactions based on folding energetics studies. However, it is not clear which is the most crucial interaction leading to major free energy change and responsible for changing from unfolded to folded states. The present chapter has discussed various factors involved in protein folding and their contribution toward folding energetics.

Chapter 4: Energy landscape theory and funnel concept have been recently introduced in finding solution of protein folding. They are being used to design various motifs present in the protein using their primary structure, prediction of order of native contact formation during folding, etc. Here protein folding is not characterized by single pathway but enormously large number of equivalent folding pathways reaching to the point of native state with single conformation of the backbone. The present chapter is based on understanding energy landscape and its significance toward prediction of three-dimensional structure of the real protein based on various topological parameters and native inter-residual contacts.

I have been introduced with the problem of protein folding by two academicians: Prof Arvind M. Kayastha (School of Biotechnology, Banaras Hindu University) and Prof. M. V. Jagannadham (Molecular Biology Unit, Banaras Hindu University). Their guidance and unflinching encouragement have generated keen interest in me to thoroughly understand the problem. I was so fascinated with the problem that I am still stuck to it and expecting to give some solution to it. Prof. Dinakar M. Salunke (Structural Biology Unit, UNESCO-Regional Centre for Biotechnology) has introduced me with all the recent tools and techniques used to find suitable solution to the problem. His involvement with his originality has triggered and nourished my intellectual maturity in the field. Further, I would like to thank my parents and hubby (Dr Yogesh K. Sharma) for their emotional and motivational support in writing this book. I have tried my best to make this book; simple and easier so that it can be approached by all the classes (students, academicians, and researchers) to understand the problem and creating enthusiasm towards finding suitable solutions.

Gurgaon, India   Alka Dwevedi

Contents

Chapter 1
Research Updates on Protein Folding

1.1 Protein Folding Problem and its Challenges

The principle guiding protein to fold into its native conformation has been searched by biochemists, structural biologists, molecular biologists, and recently by bioinformaticians. They have discovered various facts and indicated to find the solution by coming decades. There are various challenges toward them, most importantly the existence of folding intermediates for only < 1 s due to which their isolation is an unapproachable task to carry thorough analysis [1]. However, kinetic characterization of folding intermediate has been successfully done using hydrogen exchange (HX) methods in number of cases [2]. Various theoretical models have been given to elucidate protein folding pathways and their principles. Theoretical biologists are trying hard for the prediction of three-dimensional structure of native proteins from their amino acid sequence by using principle of folding pathways. It is well understood that amino acid sequence of the protein has got the code of three-dimensional structure of the native protein. The environmental factors comprising both physical (temperature, pressure) and chemical factors (salt, suitable solvent, pH, etc.) as well as presence of chaperones are the major factors which actually govern the correct protein folding. The alteration in any of these environmental characteristics can disrupt the structure *via* interference of the folding mechanism. Protein folding is the cooperative action of foldons (small unfolding/refolding units) which defines the unit steps in folding pathway. There is stepwise addition of each foldon unit during folding in a sequential manner with each step being guided by attainment of stability. Finally, native protein with maximum stability and activity is generated [3].

Understanding the mechanism of protein folding pathway would not only help in prediction of three-dimensional structure of protein but also help in curing various diseases caused by protein misfolding. Protein misfolding refers to the failure of a protein to achieve its tightly packed native conformation efficiently due to its reduced stability by environmental change or mutation. Misfolded proteins (characterized by severe aggregation and degradation) are responsible for number of incurable diseases like Alzheimer's, Parkinson's, mad cow, cystic fibrosis, etc. In case of Alzheimer's, there is aggregation of amyloid precursor protein due to its misfolding

© The Author(s) 2015
A. Dwevedi, *Protein Folding*, SpringerBriefs in Biochemistry and Molecular Biology,
DOI 10.1007/978-3-319-12592-3_1

caused by its binding to apolipoprotein E (apoE) found in the blood stream. However, it is not clear which form of protein apoE, i.e., apoE2, apoE3, or apoE4 is responsible for misfolding of Aβ [4]. Misfolding of amyloid protein leads to formation of "neuritic plaque" in the brain. The process is very slow due to which Alzheimer's disease is usually found in old patients. Another affect of protein misfolding is the mad cow disease which is transmitted from animals to human. It is caused by the misfolding of prion proteins [5]. The catalyst responsible for prion misfolding is not clear; it is believed that misfolding is self-perpetuated. The research on protein misfolding has concluded from various cases that mutation leading to instability of the protein is the major cause [6]. However, the root cause of mutation responsible for protein misfolding leading to various diseases is still not found. Lots of efforts are being made across the world to solve the problem of protein folding. The present chapter is the compilation of important research contributions in the arena of protein folding made till date. Following chapters are based on understanding the problem and its appropriate solutions.

1.2   Research Contributions

1.2.1   Before 1960s

1930

This year marked the beginning of protein denaturation using urea. Hopkins had developed a simple method of determining rate of protein denaturation. His experimentation was based on the fact that denatured protein on diluting with water would lead to precipitation of denatured product, while folded protein remains solubilized. He did complete precipitation of denatured protein (albumin) using ammonium sulfate and collected the precipitate. The precipitate was filtered, dried, and weighed and compared with total protein being used for denaturation. Thus, calculated rate of denaturation is based on calculation of total protein taken for denaturation divided by total protein denatured [7]. The method was found to be so successful that it was being adopted by number of workers across the world.

1931

Ultracentrifugation (centrifugation at very high speed) came to limelight. It was being used to determine the molecular weight of the proteins as well as their stability as function of various physical and chemical factors [8, 9]. It can only be used for globular proteins and not for the proteins with extended conformation like wool keratin. Speakman did studies on stability of proteins at different pH. He found that monodisperse proteins have wide pH stability than polydisperse protein solutions [10].

1936

Protein denaturation studies were continued using various physical and chemical factors. Fischer did heat denaturation studies on monodisperse purified serum–globulin solution. He studied denaturation using nephelometer (instrument for measuring concentration of suspended particulates in a liquid or gas colloid) and Pulfrich photometer (colorimeter used in clinical laboratories) based on the fact that denaturation leads to change in optical properties as well as change in concentration of suspended protein molecules in a solution. The denaturation was done using exposure to higher temperatures ($\leq$70 °C) leading to coagulation of protein molecules [11].

1940

This period was marked with the discussion on native state of protein and homogenous pure solution of protein. Do the proteins have same behavior (conformation) or it changes with isolation? Majority had supported that there were alteration in proteins being purified from complex mixture. It was also raised that it might be due to difficulties during separation and inconsistency in amino-acid analyses, as found in case of hen's egg-white albumin. Ultracentrifugation studies had also revealed that native and purified ovalbumin had different sedimentation constants [12].

1953

The denaturation studies on ovalbumin were carried in presence of physical and chemical factors, *viz.* hydrostatic pressure, temperature, pH, salts, detergents, sulfhydryl reagents, various organic substances, and urea. Rate of denaturation was measured using change in optical rotation. From these experiments, it was found that; there were major changes in optical rotation of protein solution in presence of high concentration of urea. Denaturation was not found to be the first order reaction; rather urea formed a complex with the protein (ovalbumin) contradictory to the view of its getting bound to localized area of protein in nonstoichiometric manner. Further, it acted by breaking hydrogen bonds within the protein to carry out denaturation similar to heat denaturation [13, 14].

1.2.2   From 1960 to 1980

In 1960s, the major question raised by the researchers was: what is the speed of protein folding? Refolding reactions of the few proteins were found to be in the seconds' time range. It might due to non-availability of ultrasensitive instruments during those days to monitor protein folding. However, now there are several examples reporting protein folding in 10^{-3} s.

1968

In 1968, Cyrus Levinthal computed rate of folding of small protein (Cytochrome *c*). He concluded that it would be longer than the lifetime of the universe by random search of all possible backbone conformations, which is impossible. Then it was raised how protein fold so fast [15]? In 1969, he noted that estimated time of protein folding was very high, that is, in astronomical numbers (10^{143}). It was due to very large number of degrees of freedom in an unfolded polypeptide chain. He explained his calculations: for example, a polypeptide of 100 residues has 99 peptide bonds leading to 198 different φ and ψ bond angles. If each of these bond angles can be in one of the three stable conformations, the protein may misfold into a maximum of 3^{198} different conformations (including any possible folding redundancy). Therefore, correctly folded protein would undergoes sequential conformations to attain native state requiring a time longer than the age of the universe. It could not be possible as there would be no cellular activities, and thus no life would have existed on earth. Later, it was suggested that protein folding could be sped up by the rapid formation of local interactions serving as nucleation points and determinant of further folding of the peptide [16]. Experimental analyses have confirmed that there are protein folding intermediates and the partially folded transition states that fasten protein folding by several folds. Various experimentalists across the world, searched for folding intermediates in proteins (small or big) [17–19]. Fast folding due to formation of folding intermediates have been reported in various cases, *viz.* Arc repressor (small dimeric protein having monomer of 53 residues) undergoes very fast refolding at the diffusion-controlled limit as found by mutating buried ionizing residues (Arg-31, Glu-36, and Arg-40). The wild-type protein refolds in a second-order reaction at low protein concentrations (16 μM), further kinetics of refolding becomes complex at higher protein concentrations. At 25 °C, the second-order refolding rate constant was 8×10^6 M^{-1} s^{-1} for wild type and 3×10^8 M^{-1} s^{-1} for the mutant. It was found that refolding rate constant for the mutant was inversely proportional to the solvent viscosity while it was independent in case of wild type. This further confirmed that refolding was diffusion-controlled reaction. It was concluded that mutations have lowered the free energy barrier from unfolded state to folded state. Further, free energy was lowered by diffusion of monomers. They had given model of protein folding as: folded microdomains diffuse together, collide, and merge to give native protein [20, 21].

1970

The three-dimensional structure of tRNA was determined by X-ray diffraction (XRD) using tRNA crystals. It was found that it has basic cloverleaf structure which was further folded into L-shaped structure as proposed by Holley [22]. tRNA folding is a two-state process as found using calorimeter. Applying this calorimetric criterion to protein folding had found that there were no detectable intermediates. It was due to the fact that protein folding was much faster than tRNA folding process, due to which intermediates so formed during protein folding were not detected by calorimetric criterion [23]. Later, came Eigen–DeMaeyer temperature (T)-jump apparatus having deadtime of less than 10 μs. It was used to detect a rapid reaction in the thermal unfolding of ribonuclease A (RNase A). They studied mechanism of folding over wide range pH (from 1 to 7). Specifically, pH 1.3 was chosen to study thermal unfolding in the range from 20 to 60 °C. Rate of reaction was found to be in the millisecond time range. Protein unfolding was measured using oscilloscope at 287 nm to monitor exposure of buried tyrosine groups. Here, they were able to detect the presence of intermediates (partially folded states) formed between 10^{-5} and 10^{-1} s. These unfolding intermediates were not detected previously by various workers. These studies supported the model based on nucleation dependent sequential folding. Further these studies had emphasized the significance of studying transient phase of sequential protein unfolding in understanding protein folding pathway [24].

1971

It was confirmed that there is at least one intermediate present during protein unfolding as found using RNase A and chymotrypsinogen A. However, there was clashing on rate of synthesis of these intermediates. It was ranged from seconds to microseconds [25]. Ikai and Tanford [26] had also found biphasic unfolding and re-folding kinetics of Cytochrome c (Cyt c). These studies were based on a quantitative treatment using guanidinium hydrochloride (GdmCl), relating kinetic amplitudes to relaxation times. These kinetic studies of the protein denaturation and renaturation had indicated that metastable intermediates were not on the direct pathway between native and denatured states. This suggested that the initial steps of protein folding involving structureless polypeptide chain may often be rapidly reversed without influence on the ultimate result. Brown and Klee [27] had reported that amino acids from 3 to 12 of RNase A, at N-terminal site, were forming partial helix at tempera-ture near $0\,°C$. Further, the chemical shift of His12 in S-peptide matches the chemi-cal shift of the transient folding intermediate of RNase A. This was an exciting result which had elucidated the mechanism of helix formation in peptides. Further, these finding gave strong support to the hierarchic model of folding.

1973

Native protein is characterized by presence of tertiary structures involved in vari-ous processes by binding to specific ligands. Garel and Baldwin [28] had designed an experiment that would confirm the presence of partly folded intermediate be-tween native and unfolded protein. They took an inhibitor (2'CMP) specific for RNase A and monitored its refolding by the absorbance change due to binding of 2'CMP. Here binding served as the probe for the formation of native state of RNase A. Further, it was also helpful in studying rate of folding and its mechanism, par-ticularly significance of nucleation process. It was found that completely unfolded RNase A using 6 M GdmCl followed both fast and slow refolding reactions leading to native protein. It was concluded that there were two different forms of unfolded RNase A; fast folding form (U_f) and slow folding form (U_s). It was not clear, why these form existed and what is the difference between them.

1975

John Brandts and coworkers had proposed a proline isomerization model to study U_f and U_s forms of RNase A. Native RNase A has two *trans* and two *cis* proline residues. It was suggested that during fast unfolding of RNase A, U_f was formed which had *cis* state while it was in *trans* state in U_s. *Trans* isomerization of proline was a very slow process which was responsible for the formation of two unfolded states. This model was very attractive, but it could not provide further evidences to support [29]. Subsequently, it was discovered that refolding of RNase A was much more complex involving a number of fast and slow folding species in the folding pathway [30, 31].

1978

Schmid had designed a refolding assays for measuring $[U_f]/[U_s]$ ratio at different times after adding $HClO_4$. His experiments confirmed that difference between U_f and U_s was due to proline isomerization only. He had also found that protein folding can be sped up by addition of stabilization salt like $(NH_4)_2SO_4$. His experiments were concluded with the facts: slow folding species (U_s) of RNase A was refolded into native-like intermediate (I_N) while fast folding species (U_f) was folded into native (N) state. Both native (N) and native-like intermediate (I_N) bind to 2'CMP with similar specificities. These experiments proved that there were structural folding intermediates present during protein folding. Therefore, it can be concluded that protein folding can proceed *via* alternative pathways supporting Levinthal's argument that folding proceeds by the fastest route if alternative pathways are available [32–34].

1.2.3   From 1980 to 2000

1980

It was raised that whether folding intermediates so formed during unfolding of protein have same kinds of secondary and loosely woven tertiary structures as found in native protein or they are different. Lesk and Chothia did some experimentation in this respect and came out with various facts. They found that approximately 50 % of total buried surface area present in protein is characterized completely by secondary structures. They suggested that secondary structures present in unfolding intermediates are part of native protein with more organized framework and complexity in interactions with their nearby main as well as side chains [35].

1983

It was found that there was non-covalent aggregation which occurred while protein followed path to fold, particularly during intermediate states. It was demonstrated in lactic dehydrogenase from pig muscle [36]. There are various sites present in this enzyme which are susceptible to fragmentation on treatment with pepsin particularly at acidic pH. They did denaturation studies of lactic dehydrogenase *via* treatment at pH 2.3 (acidic). Denatured lactic dehydrogenase was treated with pepsin at various time intervals and proteolytic products so released were analyzed by sodium dodecyl sulfate polyacrylamide gradient gels. There were distinct changes in the fragmentation pattern consisting of undigested monomers (35 kD) and 12 fragments with molecular mass ranging from 5 to 31 kD following acid denaturation of lactic dehydrogenase. There were two intermediates named M_1 and M_2 detected during enzyme denaturation. Proteolytic analysis of M_1 and M_2 had revealed that as the enzyme moves from one intermediate state to other, there was exposure of domain responsible for enhancing aggregation with first order kinetics.

1988

The nuclear magnetic resonance (NMR) pulse labeling method was applied to the folding of RNase A [37]. They found that there was an early intermediate during folding of RNase A using exchange reaction between the backbone NH protons present in the folding protein and solvent protons. There were only limited set of the NH protons present in RNase A which were available for refolding studies. Structural analysis of the intermediate using two-dimensional ^{1}H-NMR had revealed that there was formation of stable secondary structures leading to subsequent formation of the complete tertiary structure. In the same year, Englander et al. [38] did refolding studies on Cytochrome c. They emphasized on collection of detailed structural information on folding intermediates based on hydrogen exchange using two dimensional ^{1}H-NMR. They showed that there was formation of secondary structures at many defined sites along the polypeptide chain on a timescale ranging from milliseconds to minutes.

1990

This year emphasized on understanding hierarchy of interactions responsible for stabilizing the native state of protein. The best model chosen to study hierarchical interactions was folding intermediates (I). Partly folded apomyoglobin intermediate (I) was taken and characterized using two-dimensional ^{1}H-NMR by slowly trapping exchanged NH protons. It was found that subdomain of partly folded apomyoglobin contain both A helix (from the N-terminus), G and H helices (from C-terminus) supporting hierarchical model of folding. Further, these folded A, G, and H helices were not very stable [39].

1991

Hughson et al. [40] did site-directed mutagenesis to study stabilizing interactions in folding intermediates. These mutations were particularly introduced into the packing sites of $A.H$ and $G.H$ helices, aiming to destabilize native apomyoglobin (N). These mutations had no effect on intermediate I, indicating that "I" form is not stabilized by $A.H$ and $G.H$ packing interactions. There was only small increase in the stability of "I" when mutations were introduced in the side chain having high nonpolar surface area. It was suggested that "I" must be stabilized by relatively nonspecific hydrophobic interactions as indicated by nonpolar mutations. They gave the model of "molten globule" based on the fact that only a part of polypeptide being involved in formation of "I". It was also found that "I" had only 35 % of α-helical content with respect to "N" state.

1993

Barrick and Baldwin [41] made a thorough survey of the stability of the folding intermediate (I) as a function of pH and concentration of urea. They studied unfolding of sperm apomyoglobin from pH 2 to 8 and from 0 to 7.6 M urea using circular dichroism. The data presented gave two series of unfolding curves: (1) acid-induced unfolding carried out in the presence of various concentrations of urea and (2) urea-induced unfolding at various pH values. Both of the unfolding curve fitted well into the simple three-state ($U=I=N$) model. Thermodynamics parameters obtained for U, I, and N had revealed that free energy of I was closer to U than to N, indicating that side chain packing had major contribution in the stability of native protein structure. The equilibria between N and I, I and U are equally sensitive to urea, suggesting that much of the surface of I is inaccessible to solvent. The acid induced unfolding of N state of apomyoglobin was due to titration of approximately two histidine residues.

1994

The workers were busy in identifying various types of intermediates present between unfolded and native states of the protein to understand the mechanism of protein folding. They designed various experiments to change kinetics of protein folding to get different types of intermediates. Houry et al. [42] did unfolding studies on RNase A with the conditions: 1.5 M GdmCl, pH 3.0, at temperatures $\leq 15\,^{\circ}\mathrm{C}$ (double-jump experiments consisting of an unfolding step at 4.2 M GdmCl and pH 2.0 followed by a refolding step at 1.5 M GdmCl and pH 3.0 were carried out to monitor the unfolding process) to slow the refolding process. They used absorbance and fluorescence detection methods to study unfolding kinetics of RNase A. Previous reports had shown that there were, U_f and U_s species obtained from fast and slow phases of unfolding of RNase A, respectively [43, 44]. Experiments carried by Houry et al. demonstrated a new unfolded species U_{vf} and gave a new model of unfolding pathway. They proposed that U_{vf} was resulted due to isomerization of X-pro peptide bonds on unfolding on RNase A. Comparative studies on U_f, U_s, and U_{vf} have found that, U_f: Pro 114 was in *trans* conformation, U_s: Pro 93 was in *trans* conformation, U_{vf}: both Pro 114 and Pro 93 were in *cis* conformations. It was found that proline in *trans* conformation (nonnative state) had impeded refolding of RNase A due to which there was very slow unfolding/refolding in case of U_s and very fast in case of U_{vf} with both prolines in native states.

1995

Kiefhaber et al. [45] had started their work to explore structural intermediates in small proteins. Thorough analyses were done on RNase A but structural intermediates were not reported till this year. The RNase A containing β-sheet on complete unfolding opened and became accessible for proton exchange in a single cooperative step. Thus, previous workers could not detect structural intermediates as measured by hydrogen exchange with unfolding rate being measured by optical probes of tertiary structure, particularly proline. They designed an experiment to study unfolding of RNase A at very slow pace. They performed unfolding of RNase A at $10\,^{\circ}\mathrm{C}$, pH 8.0, 4.5 M GdmCl (condition when there is very slow unfolding) and measured using one-dimensional NMR. Exchange of peptide NH protons (^{2}H-^{1}H) was used to monitor structural opening of individual hydrogen bonds during unfolding. Unfolding kinetic model was developed based on hydrogen exchange. During experimentation, they took 49 protons (47 out of 49 were hydrogen bond acceptor) to carry their proton exchange studies. The results indicated that there was only one step responsible for breaking entire network of peptide bond in the polypeptide chain of native protein (RNase A). They concluded that the unfolding of RNase A also followed three-state model with the presence of unfolding intermediate (I) having free side chains. The intermediate was called as "dry molten globule". Hoeltzli and Frieden [46] had also confirmed presence of dry molten globule during unfolding of *E. coli* dihydrofolate reductase as found using ^{19}F-labeled Trp residues.

In previous years, it was confirmed that folding of apomyoglobin was proceeded *via* a molten globule intermediate (low-salt form; I_1). Three (A, G, and H) out of total eight α-helices of myoglobin were found to be present in I_1. Loh et al. [47] have found a second intermediate (I_2) which was more stable than I_1. Both intermediates were typical molten globule intermediates with loosely packed framework. They generated second intermediate (I_2) by treatment of native state (N) with trichloroacetic acid. Various properties of I_2 were similar to N state with only difference of lacking fixed side chain. They concluded that folding of apomyoglobin occurred *via* single pathway with accumulation of various intermediates of increasing stability and structural framework.

1996

There were various questions which were remained unanswered, *viz.* kinetic process of formation stable intermediates, protein folding is cooperative or noncooperative process, criteria of determination of state of folding (two states or three states) etc. Jamin and Balwin [48] had reported that folding/unfolding of apomyoglobin could be measured even in millisecond range using stopped-flow measurements of tryptophan fluorescence. Kinetics studies had revealed that formation of intermediate (I_1) during unfolding apomyoglobin was highly cooperative. However, due to lack of techniques which could measure fast-reactions adopted during unfolding of apomyoglobin, they could not give any conclusion related to folding transitions.

1997

Luo et al. [49] had tried site directed mutagenesis to know the process of cooperativity during protein folding/unfolding. From previous studies, it was established that the unfolding apomyoglobin at pH 4 produced folding intermediate containing A, G, and H helices of myoglobin. They inserted various helix destabilizing mutations, particularly in A and G helices. It was found that single glycine/proline mutations could destabilize unfolding intermediate considerably. These mutations were found to destabilize native helix propensities which in turn affected the stability of unfolding intermediate. They found that wild type stable protein (apomyoglobin) without mutations was unfolded *via* two-state reaction while unstable mutant was unfolding *via* reactions which were more complex with three-state reaction. It was concluded that stability was the key factor in determination of type of folding reaction adopted by the protein.

1.2.4   From 2000 till Date

2001

Proteins are interesting biomacromolecules with their functioning dependent on their three-dimensional conformations. Various workers tried to synthesize proteins like synthetic molecules with their ability to fold in a controlled manner [50, 51]. It is still a challenging aspect to synthesize such "foldamers" with their capacity to undergo various chemical processes and interactions. Oh et al. [52] did some work related to synthesis of oligomers from short chain segments which were ligated through an imine metathesis reaction. It was strongly believed that reaction leading to utmost stability would lead to formation of conformationally ordered sequences.

2003

Computational analyses of protein folding came into limelight with information collected from experimental data. Molecular dynamics simulations at atomic resolution have been used to get rate constants and structural details of native protein obtaining data consistent with experimental work. Here, it is possible to map folding intermediate using experimental φ-ψ values but not native protein due to limitation in atomic resolution being used (μs range) [53–55].

2005

It is believed by the scientific community that almost various aspects have explored related to protein folding but we still do not know the basic answer that how the primary sequence of protein determines its three-dimensional structure. This problem is more complex in case of membrane proteins as the field is still not intensely touched by structural and molecular biologists. Twenty-first century has marked the beginning of understanding membrane proteins and protein data bank (PDB) finds many entries of membrane proteins [56–58]. Bowie [59] has started few studies on folding/unfolding of membrane proteins, particularly emphasizing significance of distinct fold.

2007

Folded protein has the ability to bind effectively to its ligand. It has been found in most of the eukaryotic proteins present in disordered states under physiological conditions which get folded into ordered states on binding to their cellular targets. Now the question arises how binding promotes protein folding. Sugase et al. [60] have done NMR studies on phosphorylated kinase inducible activation domain (pKID) of the transcription factor cAMP response element-binding protein (CREB) forming complex with KIX domain of CREB. The C-terminal of pKID present in partially folded states becomes stabilized on binding to KIX due to enhanced hydrophobic interaction imparting stability to the pKID besides helping in complex formation. These studies will help in understanding the mechanism of various proteins having disordered structures with diverse functions.

Mok et al. [61] have found that hydrophobic collapse due to strong inter-residual contacts between side chains is responsible for formation of native protein. Here they have taken TC5b (mini-protein) to understand the mechanism of fast folding. It is also suggested that there are some unfolded and partially folded proteins known to possess biological function. Further, these are also responsible for promoting aggregation which is an important aspect to study due to its significance in various diseases like Alzheimer's, Parkinson's, etc.

2010

Shank et al. [62] have found the significance of various regions present in protein structure responsible for contributing toward rate of folding. They have used single-molecule optical tweezers approach for inducing selective unfolding of particular regions of T_4 lysozyme and monitor their effect on other regions. They have concluded their studies with the fact that there are certain structural elements present in the protein which actually controls rate of folding and it's energetics by keeping a check on folding cooperativity between its domain, and thus bring an efficient folding.

2011

There are lots of proteins being synthesized under *in vivo* conditions which are not properly folded. These misfolded proteins are continuously removed by the efficient cellular machinery. Further, cells are well equipped with chaperones playing crucial role in promoting efficient protein folding. Hartl et al. [63] have found that aging cell looses this efficiency which is the major cause of various diseases like Alzheimer's and Parkinson's. They suggested that detailed analyses of proteomics would help in understanding these pathological states.

2014

Motlagh et al. [64] have done studies on the mechanism of allosteric proteins (binding to ligand leads to generation of signals transmitted to distal functional moiety). They have done various statistical calculations of interactions which are responsible for transmission of information from one moiety to another. These analyses have given detailed allosteric mechanism including regulatory strategies being adopted by various allosteric proteins.

Chapter 2
Classes of Protein and Their Folding

2.1 Introduction

The protein is functionally active in its three-dimensional native state. Almost all cells' machinery is based on the involvement of number of proteins. Structural code of proteins is hidden into the primary structure containing sequence of amino acids. It has always been a topic of research to investigate the mechanism of determining proteins' three-dimensional details from its primary structure. Studies on protein folding have shed various important facts; *viz.* the internal core of globular protein is formed by hydrophobic amino acid residues which are held together by van der Waals forces while surface is dominated by charged and polar side chains. Native state of the protein is the most active and stable state with a specific conformation determined by polypeptide backbone and intermolecular interactions. Native state of the protein has lowest free energy due to hydrophobic and electrostatic interactions as well as hydrogen bond energy [65]. Further lowest conformational entropy due to constrains imposed by ψ and φ bonds of main and side chains [66].

Protein folding is a complex issue in case of multidomain proteins (containing more than one domain). Generally proteins with amino acids more than 200 belong to the category of multidomain. They are found to have independent folding of each domain behaving like independent whole protein. However, presence of other domains is helpful in imparting stability to the respective domain. It has been found that rate of folding is lowered when folding pathway of single isolated domain is studied. Rate limiting step in folding of multidomain protein is the interfacial interactions between individual domains [67, 68]. Folding kinetics follows an initiation fast phase followed by intermediate state lacking various properties of native protein, more labile to proteolysis and lacks catalytic activity and finally last stage with very slow rate of folding. It is during last stage when pairing of already folded domains present in the protein takes place. Rate of folding is found to be inversely related to the solvent viscosity [69]. High protein concentration as found in cellular environment has been found to be helpful in domain pairing but it usually promotes dimer formations rather monomer [70]. It has been found that ligand binding is also

A. Dwevedi, *Protein Folding,* SpringerBriefs in Biochemistry and Molecular Biology, DOI 10.1007/978-3-319-12592-3_2

one of the factor responsible for proper protein folding as found in various cases (troponin C site III, Arc repressor, Trp repressor, p53, HIV gp41) [71].

Molecular chaperones are the major proteins guiding protein folding and proper assembly of macromolecular structures in prokaryotic as well as eukaryotic cells. The most important class of chaperons is heat shock proteins, *viz.* Hsp10s, Hsp40s, Hsp60s, Hsp70s and Hsp90s. Efficient protein folding is the pre-requisite of proper functioning of cell machinery which depends on coordinated functioning of chaperones, chaperonins, and various auxiliary cofactors [72]. In some cases, this machinery fails due to which synthesized proteins could not be folded properly leading to their aggregation. The failure of a specific protein to adopt its native and active state is called "protein misfolding". Protein misfolding has wide range of pathological implications due to loss of normal cellular functions. Misfolded proteins form fibrillar aggregates called amyloid fibrils which are resistant to ubiquitin–proteasome degradation system due to which their accumulation occurs inside/outside the cell [73]. Further, aggregation and amyloid formation of a protein may also be promoted due to various point mutations which destabilize its three-dimensional structure and induce amyloidogenic conformations. Formation of the correct native-like disulfide bridges (between a pair of cysteine residues) is another factor in protein misfolding. Greater the number of disulfide linkages, higher is the chance of misfolding due to errors in disulfide pairing [74].

2.2   Structural Classes of Protein (Fig. 2.1)

(www.swissmodel.expasy.org, www.wikipedia.org/wiki/structure, www.cryst.bbk. ac.uk)

a. All-α

- *Lone helix*: Small proteins containing a single helix, *viz.* alamethicin (transmembrane voltage gated ion channel).
- *Helix-turn-helix motif*: Two helices lying antiparallel connected by short loop, *viz.* RNA binding protein Rop.
- *Four-helix bundle*: Bundle of four helices connected by three loops. Here, interfaces in between the helices are hydrophobic while surface is hydrophilic. They are present in photosynthetic reaction centre, membrane spanning region of G-protein coupled receptors, steroid-binding proteins like uteroglobin, ferritin, cytokines (interleukin-2, granulocyte-macrophage colony-stimulating factor, GM-CSF), DNA binding proteins (usually transcription factors), etc. They are also present in globin fold containing cluster of two bundles, each of four helices. Globin fold is also called "Greek key helix bundle" due to its topological similarity.
- *Helix-helix packing*: α-helices are packed in such a way that they have complementary interfacial regions with buried side chains. Their surface is not smooth, characterized by grooves and ridges with each ridge are at 26° from the main axis, for example: carboxypeptidase A, flavodoxin, subtilisin, etc.

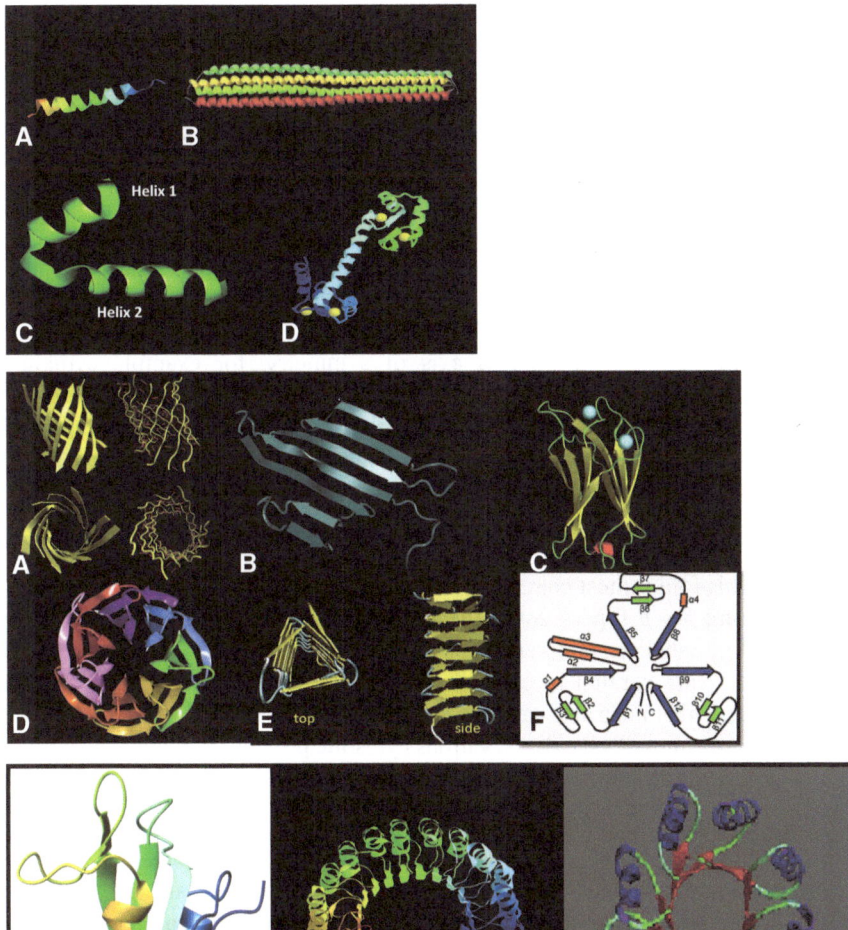

Fig. 2.1 Structural representation of protein models belonging to all-α, all-β and mixed class. **(I) all-α**, **A**: lone helix, **B**: Four-helix bundle, **C**: Helix-turn-helix motif, **D**: Helix-helix packing. **(II) all-β**, **A**: β-Barrels, **B**: Up and down antiparallel β-sheets, **C**: β-Sandwiches, **D**: β-Propellers, **E**: β-Helix, **F**: β-Trefoils. **(III) Mixed class**, **A**: $\alpha+\beta$, **B**: α/β horseshoe, C: α/β barrels. (Adapted from www.swissmodel.expasy.org, www.cryst.bbk.ac.uk)

b. All-β

- β-Sandwiches: They are also called immunoglobulin fold containing β-strands forming two sheets which are packed like sandwich. The packing of two sheets is either aligned (mean angle between the two sheets is ~30°) or orthogonal (two sheets are at 90°). The two sheets are independent which are linked by residues not in β-sheet conformation. Here side-chains are not fixed at any angles to the interface. Examples are: superoxide reductase, clathrin adaptor, transglutaminase, α-amylase inhibitor, etc.

- *β-Barrels*: Domain present in the protein containing antiparallel β-sheet without any fixed arrangement of β-strands, for example streptavadin and porin.
- *Up and down antiparallel β-sheets*: Here antiparallel β-strands making sheet are connected by loops of adjacent strand resembling to Greek key. Three up-and-down β-strands are connected by hairpins, followed by fourth strand lying adjacent to the first. Examples are: plastocyanin and γ-crystallin.
- *β-Propellers*: This fold is a superbarrel containing six, four-stranded stranded antiparallel sheets with up-down topology, for example calcium dependent *Bacillus* phytase.
- *β-Trefoils*: It has an $\sim$3-fold axis of symmetry, for example cytokinin interleukin-1β.
- *β-Helix*: This fold has resemblance to helical topology with β-strands wound round the structure, for example, monomeric left handed β-helix antifreeze protein from spruce budworm.

c. Mixed Class (α/β, $\alpha+\beta$)

- *α/β:* This fold is most commonly found in number of proteins which contains repeating β-α-β supersecondary units (right handed) with outer layer composed of α-helices and central core of parallel β-sheets. The α-helices and β-strands are parallel to each other, while α-helices are antiparallel to the strands. This fold is also called Rossman Fold named after Michael Rossman. Many enzymes of glycolysis, various cytosolic proteins, and nucleotide binding proteins have this characteristic fold.

    1. α/β horseshoe: As the name represents, they look like an open horseshoe containing a curve made by repeating units α/β with parallel β-sheet while α-helices are at the surface of the curve. The β-strands are parallel to the central axis while they are slightly slanted with respect to each other. For example, placental ribonuclease inhibitor.
    2. α/β barrels: Here sequence of eight β-α with first strand hydrogen bonded to the last strand, forming a barrel-like structure. The fold is not open rather than closed like barrel with α-helices situated on only one side of the β-sheet. The most important example: triose phosphate isomerase.

- *α+β:* They contain significant α and β secondary structural elements, not having any specific topology. Example: cysteine proteases (papain and actinidin), DNA-binding protein, microbial ribonucleases, lysozome, chalcone isomerase, ribonuclease-H, carbonic anhydrase, serine protease inhibitor, thymidylate synthase etc.

2.3   Correlation of Protein Folding with Structural Classes

Protein folding, particularly *in vitro* has been a most interesting aspect toward physicists, chemists and biologists. The ability of proteins to fold spontaneously immediately is the most crucial fundamental problem being solved since 1960s. Protein attains its native state *in vivo* with the help of various chaperones just after its synthesis on ribosome. *In vitro* protein folding has been given more priority in research with its association of various interesting facts, *viz.* how the chain can find its most stable structure within seconds, prediction of three-dimensional structure from amino-acid sequence of the protein etc. However, both *in vivo* and *in vitro* protein folding have emphasized that the native state is the most thermodynamically stable state with $\Delta G = 0$. Protein can take zillions of possible conformations, once the right stable conformation is achieved then deviation of ~ 1 Å can strongly increase the chain energy by several folds [75]. There must be some specific folding pathway which propels the unfolded protein in that particular direction making the process so fast. Molecular simulations studies using lattice models of protein chains have shown that protein folding is initiated by nucleus formation with rate of folding dependent on the size of the protein while whole process is under thermodynamic control. In earlier days, it was believed that the protein folding is proceeded by formation of nucleus by *N*-end followed by wrapping of remaining chain around nucleus which was proven incorrect in later days. Subsequently, various theories have been proposed to elucidate the mechanism of protein folding as discussed briefly here (www.wikipedia.org/protein_folding):

a. *Nucleation/growth model:* The rate-limiting step during protein folding is occurrence of nucleation (formation of smaller structural units). Once nucleation begins, it generates number of nuclei which fastens the protein folding by several folds. This model could not be fitted in various folding experiments observing folding intermediates.

b. *Diffusion–collision–adhesion model:* Protein folding is brought up by repeated diffusion and collisions of microdomains (enriched with hydrophobic clusters containing secondary structures) leading to generation of larger units. Here, rate of diffusion is the determining factor for rate of protein folding.

c. *Framework model:* It states that the protein folding is hierarchical which begins with formation of secondary structures followed by tertiary, accompanied with inter- and intra-chain interactions. Here, formations of secondary structures are the rate determining step during protein folding.

d. *Hydrophobic collapse model*: It is based on the fact that protein folding is brought up by rapid collapse of hydrophobic clusters followed by formation of secondary structures. It has not been experimentally verified whether formation of secondary structures is the initiating or intermediate step during protein folding.

e. *Jigsaw puzzle model:* This model states that different proteins have different route of folding pathway similar to the fact that there are multiple ways of solving jigsaw puzzle. This model is well suited for energy landscape view stating that native structure of protein is at global minimum while unfolded state at global maximum taking a shape of funnel with each molecule following different microscopic route from top to bottom.

f. *Nucleation-condensation model:* This model agrees with both the framework and the hydrophobic collapse mechanisms. It states that long-range as well as native hydrophobic interactions are important in the formation of transition state which imparts stability to formed-secondary structures.

Folding funnel model has gained much popularity with respect to other models for description of fast protein folding processes. Folding funnel as hierarchical folding has found that rate of protein folding is in minutes rather in astronomical numbers as predicted from Levinthal calculations. However, none of the models could explain the mechanism of protein folding. This section will be dealing with rate of protein folding with respect to different structural classes under *in vitro* conditions. The important factors are: size, amino-acid composition, chain length, native-state topology. Rate of protein folding has been considered to be an important aspect as it will give an insight into underlying folding mechanisms.

Protein unfolding under *in vitro* conditions is done using various denaturants (urea, guanidium hydrochloride), higher temperature, extreme pH, solvents etc. Based on the thermodynamics and kinetics studies of protein unfolding/refolding, monomeric and dimeric models are reported. Monomeric models comprise of native (N) and denatured (U) states present at the beginning and completion of reaction. They are of two types: two states and three-states.

1. Simplest is the two-state:

$$N \xrightleftharpoons{K_{eq}} U$$

where, K_{eq} represents the equilibrium constant of the reaction.

2. Three-state: Here native (N) protein unfolds through a partially structured intermediate (I) as shown:

$$N \xrightleftharpoons{K_1} I \xrightleftharpoons{K_2} U$$

K_1 and K_2 are the equilibrium constants of the two reactions.

Dimeric models comprise of more than one state during beginning and completion of the unfolding/refolding reactions. Various types of dimeric models are:

1. Two-state: Native dimer (N_2) end with two unfolded monomers ($2U$)

$$N_2 \xrightleftharpoons{K_{eq}} 2U$$

2. Three-state: There are two ways by which protein unfold/refold using three-states,

   – First case: Monomeric intermediate ($2I$) is populated between N_2 and $2U$

$$N_2 \underset{\longleftarrow}{\overset{K_1}{\longrightarrow}} 2I \underset{\longleftarrow}{\overset{K_2}{\longrightarrow}} 2U$$

- Second case: Dimeric intermediate (I_2) is populated between N_2 and $2U$

$$N_2 \underset{\longleftarrow}{\overset{K_1}{\longrightarrow}} I_2 \underset{\longleftarrow}{\overset{K_2}{\longrightarrow}} 2U$$

K_1 and K_2 are the equilibrium constants for each transition.

3. Multiple state: It exists when there are different forms of intermediates between states N_2 and $2U$. Following are the equations describing various cases.

$$N_2 \underset{\longleftarrow}{\overset{K_1}{\longrightarrow}} 2I^x \underset{\longleftarrow}{\overset{K_2}{\longrightarrow}} 2I^y \underset{\longleftarrow}{\overset{K_3}{\longrightarrow}} 2U$$

$$N_2 \underset{\longleftarrow}{\overset{K_1}{\longrightarrow}} I_2 \underset{\longleftarrow}{\overset{K_2}{\longrightarrow}} 2I \underset{\longleftarrow}{\overset{K_3}{\longrightarrow}} 2U$$

$$N_2 \underset{\longleftarrow}{\overset{K_1}{\longrightarrow}} I_2^x \underset{\longleftarrow}{\overset{K_2}{\longrightarrow}} I_2^y \underset{\longleftarrow}{\overset{K_3}{\longrightarrow}} 2U$$

Dimeric intermediates (I_2, I_2^x, I_2^y), monomeric intermediates ($2I$, $2I^x$, $2I^y$), unfolded monomers (2U), and K_1, K_2 and K_3 are the equilibrium constants for the three transitions.

It has been found that rate of unfolding/refolding depends on various factors: native-state topology [*viz.* all-α, all-β, mixed class ($\alpha+\beta$ or α/β)], size, and amino acid composition [76, 77]. As the studies on protein unfolding/refolding are progressing, various facts have been elucidated regarding correlation of rate of unfolding/refolding with various factors including structural class of proteins. However, these studies are mostly limited for two-state monomeric models with very few reports on three-state monomeric models. There are no reports on any dimeric models till date due to complexity of the reactions. Following discussion will be on correlation of folding rate with respect to two-state monomeric models only.

Brief description of the terms are given below which will be used in following text to understand the correlation of protein folding rate with various factors.

- Contact order (CO): measure of inter-amino acid contacts in the native state of protein structure. It is estimated as the average sequence distance between residues forming native contacts within the folded protein divided by the total length of the protein. Higher the value of contact orders greater would be the time of protein folding.
- Relative contact order (RCO): measure of the relative interactions of local *versus* non-local noncovalent interactions.
- Absolute contact order (ACO): average sequence separation of contacting residues.

- Long range order (LRO): measure of amino acid interactions with the distance away from more than four amino acids.

Structure topology is the main determinant of the folding rate of small proteins following two-state kinetics. They are independent of chain length as found from experimental and theoretical reports [78–80]. In case of larger proteins with observable intermediates (monomeric models), chain length is the main determinant of their folding rates. It has been found that there is no correlation of three-state kinetics following proteins with RCO while it is logarithmically related in two-state kinetics following proteins [81]. The analysis of 56 non-redundant two-state proteins with chain length varying from 16 residues (the C-terminal α-hairpin peptide of the B1 domain of protein G) to 322 residues (4 α-helix bundle of the VlsE antigen protein (PDB ID: 1L8W) was done to determine the correlation of folding rate with factors (chain length, amino-acid composition and surface topology) [78].

Before going into the detailed analyses of the report, following are given relation of RCO and LCO with factors [82]:

$$RCO = \frac{1}{N_c L} \sum_{\substack{contacting \\ atomsi,j}} d_{ij}$$

where,

L chain length
N_c total number of contacting atoms (using a 6 Å distance threshold)
d_{ij} number of residues separating those two residues to which atoms i and j belong.

$$LRO = \frac{1}{L} \sum_{\substack{contacting \\ residuesi,j}} n_{ij},$$

where, n is the number of residues separating those two residues to which atoms i and j belong with their separation always less than 12 [83].

Absolute contact order (ACO) is correlated to RCO as:

$$ACO = L \times RCO,$$

Further, two residues are considered to be in contact if the closest distance between their C_α atoms is ≤ 8 Å.

Figures 2.2a, b, c and d have shown the correlation of between logarithmic folding rates, $\ln k_f$ and basic structural and topological parameters for proteins belonging to three structural classes (all-α, all-β and mixed class). Figure 2.2a, shows correlation of total chain length (L) and rate of folding ($\ln k_f$). Straight line was observed in case of all-α and all-β with correlation coefficients as -0.80 at P-value of 5.7×10^{-4} and -0.80 at P-value of 6.0×10^{-5}, respectively. In case of mixed-class

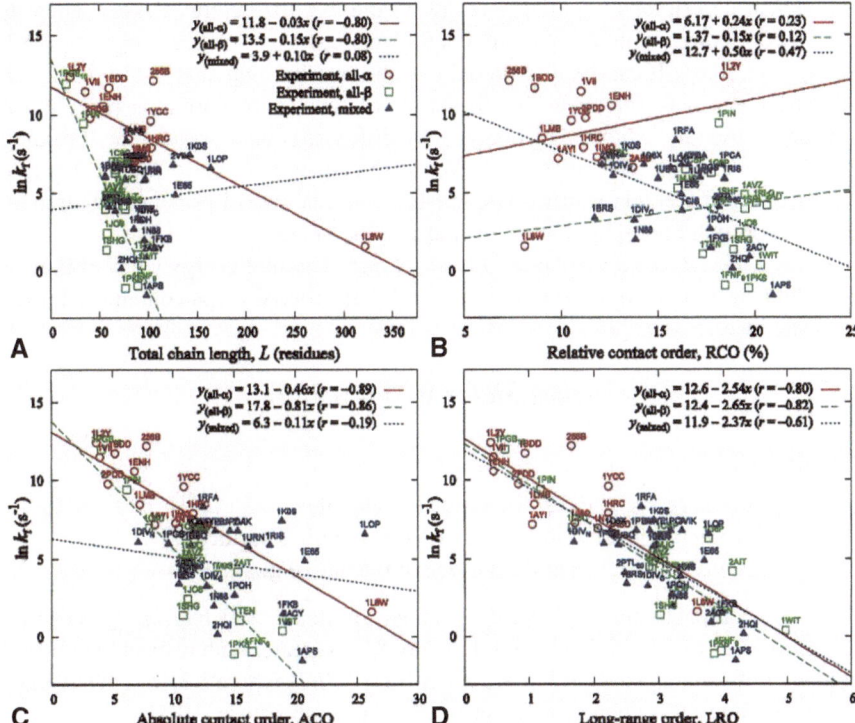

Fig. 2.2 Correlation between natural logarithmic folding rates ($\ln k_f$) of different proteins (mentioned with their PDBID) with their basic structural and topological parameters. (**A**): $\ln k_f$ *versus* protein chain length, L, (**B**): $\ln k_f$ *versus* relative contact order, RCO, (**C**): $\ln k_f$ *versus* absolute contact order, ACO, (**D**): $\ln k_f$ *versus* long-range order, LRO. (Adapted from [78])

proteins, there is almost no correlation between chain length and rate of folding ($\ln k_f$), showing that their folding mechanism is different and more complex than pure class. These results are concluded with the fact that the chain length is linearly correlated to folding rates in case of all-α and all-β two-state proteins while it's independent in mixed class. Figures 2.2b, c and d, show the correlation of RCO, ACO, and LRO with rate of folding in pure class (all-α and all-β) and mixed class. The RCO has significant impact on $\ln k_f$ for small two-state proteins of all three structural classes in the range when chain length was ~ 100–250 residues. However, it does not fit well when protein is too small, *viz.* all-α class, 20-residues [Trp-cage miniprotein construct, TC5b (PDBID: 1L2Y) of the 322-residue VlsE protein (PDB: 1L8W)]; all-β proteins, 16-residues [C-terminal β-hairpin of protein G of the 34-residue subdomain of peptidyl-prolyl *cis-trans* isomerase (PDBID: 1PIN)] with completely disrupted correlations of RCO with their folding rate. On the other hand, ACO shows strong correlation of high statistical significance with the folding rates of pure class indicating that both chain length and structural topology have significant affect on folding rate of two-state proteins. LRO has been found to be uniform topological descriptor of folding rates of all the three structural classes. There is strong correlation with correlation coefficients as: $r_{all-\alpha}=-0.80$, $r_{all-\beta}=-0.82$, and $r_{mixed}=-0.61$. It can be interpreted that the rate-limiting step is the formation of β-sheet and loop structures (i.e., formation of contacts that are long range in sequence, whose rate is limited by cooperative diffusion) due to which rate of folding is slower in case of all-β and mixed class with respect all-α class. Higher the rate of formation of secondary structures, greater would be rate of protein folding in all cases supporting hierarchical view of folding [78, 84, 85].

   Correlation of free energy with rate of folding: The free energy (ΔG) of the native globular protein which is not covered with disordered loops containing L residues can be related as [86]:

$$\sigma = 2.3RT \text{ x } 0.33 \approx 0.7RT$$

where,

g   free energy of one residue inside the globule
σ free energy lost by one residue on the globule's surface
$B_L L^{2/3}$ number of residues at the surface of the native globule

For the compact and spherical globular protein, $\sigma \approx 1/6 \times 1.2L^{2/3} \times 5RT+4/6-1.2L^{2/3} \times 0.7RT \approx 1.5L^{2/3}RT$. which is only 7.6–8.7% greater with respect to two-fold oblong or oblate ellipsoid, and $\sigma = 2.3RTx0.33 \approx 0.7RT$, where $2.3RT$ is the average residue's energy lost upon protein denaturation at temperature T [87].

   Protein folded to its native states leads to $\Delta G=0$, thus $g=-sB_L L^{-1/3}$. This suggests that the surface stability is an important parameter for achieving stable native state due to its direct interactions with the surrounding solvent responsible for number of conformational variations in the main as well as side chain. Further, if protein folding goes *via* spherical (the least unstable) intermediate structures, free energy of the fastest pathway is given as [88]:

$$\Delta G_\# = \frac{4\sigma B_{sph} L^{2/3}(B_{sph}/B_L)^2}{27}$$

whereas, the folding nucleus size (central region of the protein) given as:

$$L_\# = \frac{8L(B_{sph}/B_L)^3}{27}$$

For a spherical central globule, $B_L = B_{sph}$ and $L_\#/L \approx 0.30$, free energy is given as:

$$\Delta G_\# / RT \approx \frac{L^{2/3}}{2}$$

The maximal estimate of free energy of the nucleus corresponds to the case when the nucleus is covered with closed loops protruding from the folded into unfolded phase.

$$\Delta G_{\#} = n_{loop}\,\sigma_{loop} + n_{free}\sigma,$$

$$\sigma_{loop=}\dfrac{5RT\left[2 + \ln 2 - \dfrac{\ln L}{L^{1/2}-1}\right]}{2},$$

$$\sigma \approx 1/6 \times 1.2L^{2/3}\,5RT + 4/6 - 1.2L^{2/3}\,0.7RT \approx 1.5L^{2/3}RT.$$

The maximum free energy could be achieved at L tending to infinity to $\approx 6.7RT$, with real-size protein (with $L \approx 100$) and $\sigma_{loop} \approx 5RT$ [89].

High free energy of a loop (σ_{loop}) suggests that the nucleus used for folding is covered with loops on only one side which separates already natively folded phase from disordered one. The optimal (minimal) estimate of the maximal size of the interface between the folded and unfolded (loop containing) part of the protein is given by the largest (central) cross-section of the L-residue sphere, i.e., this border contains $L^{2/3 \times}\,(36\pi)^{1/3}/44 \approx 1.2L^{2/3}$ residues. In this case, the folding nucleus looks like a half of the native globule. At $\Delta G = 0$ for the whole native protein, free energy of this "half-globule" folding nucleus is determined only by its interface with the unfolded part of the chain.

As a result,

$$\Delta G_{\#}\,/\,RT \approx \dfrac{3L^{2/3}}{2}$$

Correlation of free energy and folding rate are given as:

$$k_f \leq k_f^{step}\,\exp\left[-\dfrac{1}{2}(L^{2/3} - 1) - 0.3\dfrac{\Delta G}{RT}\right]$$

$$k_f \leq k_f^{step}\,\exp\left[-\dfrac{3}{2}(L^{2/3} - 1) - 0.5\dfrac{\Delta G}{RT}\right]$$

Therefore, it can be stated that ΔG increases with the folding rate by a factor of $\approx \exp(-0.4\Delta G/RT)$. Rate of folding of globular protein under natural biological conditions should always be $> 10^{-2}$–10^{-3} s^{-1} to complete folding process within minutes. Rate of protein folding (k_f) decreases with the size of the protein (L) and increases with stability ($-\Delta G/RT$) of its native fold; but does not hold true for various cases. The globular protein with $L < 90$ residues having $\Delta G/RT = -8.0$ can find its most stable fold within a biologically reasonable time. The larger proteins are under kinetic control having native folds stable for reasonable time period. They have low RCO which suggests that the fold present in these proteins have least number of long loops which are responsible for lowering of folding rate. In case of small proteins (>100 residues) having much larger space (higher RCO) with negligible entropy have high rate of folding by neglecting probability of possible loop knotting, as also being experimentally confirmed. In case of very large proteins ($L>500$ residues) having $\Delta G/RT \approx -20$ with stable quasi-spherical domains, cannot fold within a biologically reasonable time unless they are divided into domains with consideration of elongated shape rather globular shape [90–92].

2.4 Protein Misfolding and Aggregation

Native states of the proteins are most stable and physiologically active state. Proper folding and maintenance of three-dimensional structure inside the cell is due to action of chaperone, chaperonin and various auxiliary cofactors. However, under certain circumstances this machinery fails and protein undergoes incorrect folding, i. e. protein misfolding. Protein misfolding is characterized by severe aggregation which is termed as amyloid. Amyloid deposition is associated with more than 20 human degenerative diseases in various parts of body. Neurodegenerative diseases like Alzheimer's, Parkinson's, Huntington's, spongiform encephalopathy which affect central nervous system. In non-neuropathic localized amyloidoses, protein deposition occurs in a certain type of tissue such as Langerhans' islands in type II diabetes. In systemic amyloidoses such as AL amyloidosis, involving deposition of immunoglobulin light-chain fragments not limited to single tissue [93]. These amyloid fibrils are rich in β-structural contents with β-strands being oriented perpendicularly to the fibril axis as shown by X-ray diffraction (XRD) [94]. Electron microscopy (EM) and atomic force microscopy (AFM) have found that each protofilament of fibril can be straight or curved with 2–5 nm in diameter. Fibrils usually consist of 2–6 protofilaments, twisting together in the form of rope or ribbon having diameter of 7–15 nm [95]. However, there are no reports on high-resolution structural details of amyloid fibrils due to limitation in spectroscopy (due to insolubility) and XRD (unidirectional crystal growth). Solid-state NMR spectroscopy has shown some hope in obtaining high-resolution data of amyloid structure. It has been found that the morphology of amyloid fibrils grown *in vitro* depends on solution conditions (buffer composition, pH, temperature, and protein concentration). The kinetics of amyloid formation is characterized by lag phase followed by log phase and stationary phase. Lag phase can be modulated by repeatedly removing aggregates or addition of aggregates as this is the phase of nucleation requiring small nuclei (i.e. aggregates) to start the process of subsequent aggregation and amyloid formation [96]. Amyloid formation is usually a nucleation-dependent reaction consisting of initial lag phase, followed by rapid growth. Presence of small nuclei of amyloid is enough to promote aggregation of complete protein solution. The most important aspect is that any protein is able to form amyloid under *in vitro* conditions particularly when environmental factors (buffer composition, pH, temperature, and protein concentration) are unfavorable [97]. However, there is variation in the morphology (*viz*. disordered aggregates, oligomers, spherical aggregates, prefibrillar aggregates) of amyloid fibrils grown under different environmental conditions. Further, there are certain sequences which are more prone in amyloid formation having higher propensity of forming β-sheet structure than α-helix. Presence of destabilizing mutations is also known for promoting amyloid formation brought up by some error in transfer of information from DNA to protein [98]. Further, alternating patterns of polar and non-polar residues promoting β-sheet formation is a potential candidate in the amyloid formation [99].The hydrophobicity of the peptide chain has been shown to influence its aggregation propensity. Further, charge of the polypeptide chain is one of the crucial promoting factors of protein aggregation. It has been found that high net charge either positive or negative prevent aggregation of the polypeptide. It has been found that β_2-microglobulin known to make amyloid fibrils under *in vivo* conditions is failed to form them under *in vitro* conditions at physiological pH suggesting that there is involvement of unknown factors inside the cell. However, presence of collagen, apolipoprotein-E, heparin, serum amyloid P component, and low concentrations of sodium dodecyl sulfate in the protein solution of β_2-microglobulin have led to its aggregation under physiological conditions [100]. Therefore, it would be an interesting study to find out various factors responsible for protein aggregation which would be helpful in giving effective therapeutics in coming future.

2.5   Protein Folding Inside Cell

Molecular chaperones are known to assist protein folding inside all types of cell. They are not only involved in proper folding of the synthesized protein but also removes unfolded or partially folded protein from cell. They are present in mitochondria and endoplasmic reticulum besides cytoplasm which clarifies that they are crucial machinery involved in proper folding of protein wherever protein is synthesized in the cell. Chaperones found in prokaryotes are less in number with least redundancy as compare to eukaryotes [72]. Action of any chaperone inside the cell works in coordination with favorable environmental conditions. Chaperones are also involved in various other functions besides catalyzing protein folding, *viz*. heat shock proteins in stress, starvation, toxicity, controlling cell cycle etc [101]. Chaperones found in nucleolus are called nucleolar chaperones, for example nucleolar multitasking proteins (NoMP's). They play an important role in the cell organization to complete various biological tasks necessary for cell survival [102].

Following are the list of well characterized chaperones working sequentially to bring proper folding of the synthesized polypeptide [103, 104]:

a. *Trigger factor:* First chaperone which interact with the nascent chain synthesized on ribosome by binding to its hydrophobic region. Further, they are also responsible for increasing efficiency of the other chaperone which is known for proper folding of the polypeptide. Trigger factor is released as soon as protein adopts its native state with the burial of the hydrophobic portions.

b. *GroEL and GroES*: GroEL (60 kDa) and GroES (10 kDa) are bacterial chaperones with their three dimensional structure resembling like ring with empty centre into which unfolded protein gets fitted. They work coordinately to assist proper folding of the protein which is brought by their conformational change forcing the protein to adopt native state while whole the process is ATP dependent.

c. *Hsp70 and Hsp90*: Heat shock proteins; Hsp70 and Hsp90 work concordantly to bring proper folding of protein. Hsp70 has three major domains (*N*-terminal ATPase domain, substrate binding domain, and *C*-terminal domain). The *N*-terminal domain plays crucial role in binding and hydrolyzing ATP, the substrate binding domain bind to ~seven residues (neutral or hydrophobic amino acid residues) of the nascent polypeptide while the *C*-terminal domain acts as a lid which covers the substrate binding domain containing nascent polypeptide. This lid is open when Hsp70 is bound to ATP bound, while it is closed when Hsp70 is bound to ADP.

Chaperone Hsp90 works independently without need of ATP binding to undergo conformational change in folding polypeptide. However, ATP is required only for stabilization of the complex (Hsp90 and polypeptide). X-ray studies have found that this chaperone has different conformations in different species *viz*. human, yeast and bacteria. This chaperone has significant role in cell's homeostasis in addition to its involvement in protein folding. It has been found that Hsp90 is being used by cancerous cells for its survival. Therefore, it is one of the targets for drug development against cancer.

d. *YidC, Alb3, Oxa1*: YidC is present in Gram-negative and Gram-positive bacteria, Oxa1 in the mitochondrial inner membrane while Alb3 is present in the membrane of the thylakoid inside the chloroplast. These chaperones help in the insertion of properly folded protein into the plasma membrane. These chaperones work by interacting hydrophobic region of protein and making these regions assessable to membrane for effective insertion.

e. *Small heat shock proteins (sHSPs) and α-crystallins (αCs)*: They are the product of housekeeping genes, thus present continuously inside the cell. They are found in almost all cell types of eukaryotes and prokaryotes. It has been found that their concentration is increased drastically on exposure of any type of stress to the cell. Their mode of action is ATP-dependent to bring proper folding of protein. These chaperones remove denatured as well as aggregated proteins, thus act as checkpoint for accumulation of only properly folded protein in the cell. sHSPs and αCs have molecular weights ranging from ~12 to 42 kDa. They have no specificity for any particular region of protein; however both preferred to bind to unfolded protein or partially folded protein. It can be said that their binding is conformation dependent rather on hydrophobicity of the segment.

Chapter 3
Energetics of Protein Folding and Significance of Intermediates

3.1 Introduction

Protein folding has now been accepted to be a self assembly process with favorable environmental parameters. The process involves free energy change from unfolded to folded state due to various interactions (ionic, covalent, and hydrogen bonds, hydrophobic interactions) leading to lowest energy state of native three-dimensional structure [65, 66]. From past decades, researchers are trying to predict native tertiary structure of the protein from amino acid sequence using folding energetics parameters. Further, efforts are also being made to understand protein–protein and protein–ligand interactions based on folding energetics studies. These studies are also helpful in understanding diverse cellular processes; *viz.* transfer of information by the ligand through specific receptor toward cell's interior, generation of antibodies against specific antigens, etc. [105]. Most of the methods based on prediction of three-dimensional structure of native protein have used calculations derived from enthalpy changes rather than free energy. This is due to inaccuracy in getting values of free energy due to number of conformational changes brought during protein folding [106]. There are number of interactions which play pivotal role during protein folding. However, it is not clear which is the most crucial interaction leading to major free energy change and responsible for changing from unfolded to folded states. Proteins are synthesized as unfolded chains which are subsequently folded into their native three-dimensional structures inside the cell. Peptide H-bond plays an important role as it marks the beginning of protein folding. However, hydrophobic factor arising from van der Waals (packing or dispersion) interactions actually guide the route of protein folding. Energetics studies on two factors have revealed that peptide H-bonds give only marginal stability with values of free energy change far less than hydrophobic interactions [107]. The other auxiliary factors in protein folding energetics are electrostatic interactions among ionized side-chains, α-helix and β-sheet propensities, and specific pairwise interactions among side-chains (*viz.* hydrogen bonds and salt bridges) [108]. The present chapter has discussed various factors involved in protein folding and their contribution toward folding energetics. Here, the change in backbone conformational entropy upon folding has also

A. Dwevedi, *Protein Folding,* SpringerBriefs in Biochemistry and Molecular Biology, DOI 10.1007/978-3-319-12592-3_3

been discussed thoroughly in addition to correlation of hydrophobic factor toward entropy and enthalpy change during protein unfolding/refolding.

3.2   Folding Mechanism and Kinetics

Proteins maintain their native structure under physiological conditions due to favorable enthalpy change *via* solvent–protein interactions exceeding magnitude of unfavorable entropy of the reaction. It is clear that protein is most stable in its native state while most unstable in its denatured state. However, the Gibbs free-energy difference (ΔG) between the biologically active and denatured states of the proteins is very small. The Gibbs free energy is the measure of protein stability; lowest value corresponds to maximum stability of protein. As the native state of the protein has lowest Gibbs free energy indicating its maximum stability with respect to any states [unfolded (U) or intermediate (I) states]. Protein stability depends on the solvent–solvent, protein–solvent, and protein–protein interactions. Further, they are stable only within a narrow range of conditions (*viz.* temperature, pressure, salts, and pH) [109]. ΔG is related to enthalpy change (ΔH) and entropy change (ΔS) during protein folding/unfolding as follows:

$$\Delta G = \Delta H - T \Delta S$$

Folding of completely denatured protein containing free-swinging peptide chain is controlled by number of physical forces having positive or negative impact on the folded chain. The entropy of the protein decreases on its folding as the polypeptide chain is arranged into ordered structure. However, entropic calculations are not only with respect to protein but also with respect to surroundings of the protein. For example, water molecules are ordered as hydration spheres around exposed hydrophobic residues of unfolded protein which increase the entropy of the surrounding. During protein folding, initially exposed hydrophobic residues to the aqueous environment are buried inside the protein disturbing the hydration order made by water molecules. Thus, entropy of the surrounding increases which overcomes the effect of decrease in entropy by folding protein leading to an overall increase in entropy. According to above relation, "$- T \Delta S$" becomes more positive due to increase in overall change in entropy of the protein. Further, enthalpic contributions due to hydrogen bonding, ionic salt bridges, and van der Waals forces subside "$- T \Delta S$" making total free energy change as negative. Therefore, protein folding is the function of internal as well as external factors responsible for making whole process energetically favorable corresponding to negative free energy.

Temperature is one of the important parameter guiding protein folding. During protein unfolding, ΔH and ΔS are dependent on temperature through the heat capacity difference, as [110]:

$$\Delta H = \Delta H_R + \Delta C_p (T - T_R) \quad \text{and} \quad \Delta S = \Delta S_R + \Delta C_p \ln(T / T_R)$$

where, ΔH_R and ΔS_R are the enthalpy and entropy changes at the reference temperature, T_R while T is the absolute temperature in K.

Structural changes during protein folding/unfolding are studied using fluorescence, phosphorescence, circular dichroism, infrared spectroscopy, nuclear magnetic resonance, and mass spectroscopy. These help in defining kinetic as well the thermodynamic parameters during protein folding/unfolding. The thermodynamic parameters (Gibbs free energy, entropy, and enthalpy) are used to define the distance between the native and actual states of protein, while kinetic parameter measures the time needed for the protein to reach the native state from a given starting state. In the native state, many of the protein functional groups are buried in the core, from which solvent water is excluded. Additionally, the amino acid side chains in the core are essentially fixed by specific interactions and tight packing. The peptide backbone is also relatively fixed throughout the protein. The amino acid side chains on the surface of the protein are also likely to have their motion somewhat restricted because of their close proximity to each other; although, this restriction should be much less than that of the interior side chains. On unfolding, the interactions between protein groups within the core are disrupted and replaced with interactions with the solvent. Also, the backbone and side chains become much more mobile and gain configurational entropy which strongly favors the unfolded state [111].

Brief description of primary forces responsible for protein folding leading to stable and active native state [www.wikipedia.org/wiki/Protein_folding, 112]:

a. Van der Waals interactions: Also called London dispersion forces, result from transient dipoles of nonbonded atoms induced in each other. Such interactions are fairly weak, short range and strongly distance dependent. They are important determining factor during protein stabilization in its native state due to their involvement in packing of atoms inside protein core as well as their interactions with surrounding solvent.

b. Hydrogen -bonds: They are noncovalent interactions that arise from the partial sharing of a hydrogen atom between a hydrogen bond donor group, such as a hydroxyl –OH or an –NH, and a hydrogen bond acceptor atom, such as oxygen or nitrogen. Many potential hydrogen bond donor and acceptor groups are present in proteins, namely the peptide backbone groups and polar amino acid side chains.

c. Electrostatic interactions: They occur between charges on protein groups. Such charges are present at the amino- and carboxy-termini and on many ionizable side chains. Charges buried inside the protein core interact strongly due to low dielectric medium in the protein's interior, which suggest that these interactions have minor role in protein stability since they are overly common. Electrostatic repulsion may be more important, not only in destabilizing the native state but also in terms of its effect on the degree of extension of the unfolded state.

d. Configurational entropy: This destabilizes the native state of the protein. The gain in the configurational entropy relates to the increased degrees of freedom available to the protein chain in the unfolded state relative to the native state. The gain comes from both the side chains and the backbone. Although the peptide backbone of most residues in a globular protein is relatively fixed, those residues that are most buried within the core of the protein have even fewer backbone degrees of freedom. The entropic effect of burying side chains is more pronounced since they have considerable flexibility on the protein surface. As larger proteins bury most of their side chains, they will have an overall larger configurational entropy change per residue. This effect may help to set a limit on the size of a globular folding domain. Further, amino acid composition also affects the configurational entropy. For example, proteins containing a large proportion of proline residues will have lower entropy in the unfolded state and thus will be more stable. The opposite will be true for proteins containing a large proportion of glycine.

e. Role of water: Water plays crucial role in the stabilization of proteins. The small molecular size of water relative to other liquids, along with its complex hydrogen-bonded structure, makes it an excellent solvent for many functional groups.

3.3   Role of Enthalpy and Entropy in Protein Folding

Rate of protein folding is governed by the extent of cooperativity among structural units guided by attainment of minimum free energy. It is applicable for both globular and elongated proteins. The underlying structural and thermodynamic mechanisms of cooperativity, particularly their energetics and extent of entropy change, have remained elusive. Further, three-dimensional structure of protein can be accurately predicted if complete information on folding energetics is known. Most of the methods for structure prediction are based on finding of minimum enthalpy of the structure rather than minimum free energy due to its easier reliable calculations [113]. There are various important protein folding energetics factors: peptide H-bonds, electrostatic interactions, α-helix and β-sheet propensities, specific pairwise interactions among side-chains, salt bridges, and backbone conformational entropy. Important parameters of protein folding energetics are given below [114]:

a. *Peptide solvation and desolvation*: Peptide desolvation is an exchange reaction accompanied by H-bonding of the buried peptide NH and CO groups. It is one of the most important energetic factors during protein folding. In case of peptide solvation, H-bonds are formed between water and peptide (NH and CO groups) which is important in protein unfolding. Titration calorimetry is used to calculate $\Delta H°$ or $\Delta G°$ for peptide solvation and desolvation. The obtained values are helpful in determining their significance in protein folding and connecting propensities of different amino acids in forming secondary structures.

b. *Hydrophobicity:* Hydrophobicity is the ability of the solute to get readily soluble in nonaqueous solvents. Thermodynamic analysis of protein solvation/desolvation of nonpolar groups in the solvent has revealed various facts. They are two-step process, involving arrangement of solvent molecules in the form of cavity followed by entrance of protein into the cavity *via* formation of van der Waals interactions between the solute and the solvent. In case of water as solvent, it makes largest contribution in making cavity due to its small size with other solvents. Further, H-bonding is another important factor towards making water as important solvent for solvation.

c. *Backbone conformational entropy:* Folded protein has single backbone conformation corresponding to native state resulting from fine balance between numbers of forces. Native state of the protein is formed by opposing conformational entropy using covalent and noncovalent interactions. Conformational entropy calculations are often problematic; however Makhatadze and Privalov [115] found that the conformational entropy change for myoglobin at 298 K is ~ 2.4 kcal K^{-1} mol^{-1}. On per residue basis, this value lies between the ΔS of sublimation and the ΔS of vaporization for organic molecules. The conformational entropy change (ΔS_{res}) can be calculated using following relation:

$$\Delta S_{res} = R \ln W$$

Where, R is the gas constant and W is the ratio of degrees of freedom between final and initial states.

Myoglobin contains 745 effective dihedral angles, each of which has up to three degrees of freedom. The upper limit of the conformational entropy change can be calculated by assuming that all rotational degrees of freedom are lost in the native state and available in the denatured state. Thus, $\Delta S_{res} = 745\ R \ln(3) = 1.6$ kcal K^{-1} mol^{-1}. The actual value in fact would be considerably smaller, since the approximation of zero degrees of freedom in the native state is too stringent for buried residues than the residues present at the protein's surface.

d. *Coulombic interactions:* Ionizing residues present in globular proteins which are partly or fully exposed to solvent are either basic (Lys, Arg, His) or acidic (Asp, Glu, Tyr). They become important at high or low pH when these residues are present in their ionized forms. A plot of temperature *versus* pH to determine T_m (temperature midpoint of thermal unfolding) has indicated sharp drop in T_m as the pH decreases from pH 7 to 2. pK values of the unfolded protein can be estimated from pK data of model peptides. However, this procedure is not suited for all cases particularly in case of large proteins with > 100 residues. One reason for the wide interest in Coulombic interactions is their efficacy in folding stability. Further, they are long range forces based on charge-charge interactions which are established at early stages in the protein folding process.

e. Salt Bridges: They are also important contributor of protein stability during folding, being formed between pairs of side-chains. However, they are least significant in small proteins than large oligomeric proteins. Proteins from thermophiles are known to have number of salt bridges.

Based on significance of these energetics factors and their correlation with protein folding, two models have been developed as briefly described below:

1. *Kauzmann's model* [116]: According to this model, the hydrophobic factor is the most crucial energetic factor during protein folding. Further, free energy change during protein folding involving burial of non-polar side chain can be experimentally determined by:
   - Identification of solutes responsible for transferring free energies of nonpolar protein side-chains
   - Determination of exact non-aqueous solvent for free energy change of hydrophobic core of a protein
   - Determination of van der Waals interactions between side chains, buried inside proteins

Quantification of hydrophobic factor in protein folding can be done by determination of surface area of a nonpolar region of the protein. According to Lee and Richards algorithm [117]. The water-accessible surface area (*ASA*) of a molecule is related to free energy change (ΔG_{hyd}) as:

$$\Delta G_{hyd} = k_h(ASA)$$

where, k_h is the proportionality coefficient
Further, ΔG_{hyd} in a protein folding reaction can be calculated using following equation:

$$\Delta G_{hyd} = k_h[(ASA)_N - (ASA)_U]$$

where *(ASA)*$_N$ and *(ASA)*$_U$ are the surface area of native (*N*) and unfolded (*U*) forms of the protein, respectively. Values of k_h varies with type of nonpolar solvent and solute used to model protein interior and amino acid side-chains, respectively. Further, the correlation of ratio of solute/solvent molar volumes also has significant affect on protein folding.

2. *Packing–desolvation model* [118]: Here the focus was on the apolar (*ap*) side-chains in the hydrophobic core which are closely packed by the van derWaals interactions. Polar (*p*) peptide groups have an important energetic role in the solvation as well as desolvation responsible for making or breaking peptide H-bonds respectively. Relating protein unfolding in aqueous solution (*N→U*) (aq) to the (hypothetical) unfolding in vacuum (*N→U*)(vac) involving desolvation of *N*, $-\Delta H_N(solv)$, and the solvation of anhydrous *U*, $\Delta H_U(solv)$. Further, change in enthalpies in aqueous and solution can be related as:

$$\Delta H_{NU}(vac) = \Delta H_{NU}(aq) + \Delta H_{NU}(solv)$$

$$\Delta H_{NU}(solv) = \Delta H_N(solv) - \Delta H_U(solv)$$

$$\Delta H_{solv} = \Delta H_p(solv) + \Delta H_{ap}(solv)$$

The unfolding of protein in vacuum involves breaking of van der Waals, H-bond and electrostatic interactions arising from difference in the dipole-dipole interactions present in the peptide backbones of native and unfolded states of protein. The van der Waals forces are well understood in model systems (5–15 peptides) but it is not yet possible to test them in proteins. However, it is useful in developing an empirical calibration expressing contribution of polar and non-polar groups in the enthalpies of protein unfolding. In contrast to Kauzmann's model, packing–desolvation model has not gained much focus due to its inadequacy in determination of energies contributed by van der Waals interactions during protein folding.

3.4 Energetics of Intermediates Formation in Protein Folding and Their Stabilization

Stability of protein has been given utmost priority due to the fact that most stable state corresponds to the most active state of the protein. Therefore, methods are being explored for stabilization strategies to be implicated at laboratory as well as industrial scale. Model proteins which have been used for exploring stabilization strategies are proteins with two-state folding equilibrium. The large proteins having complex folding equilibria involving several intermediates between unfolded and folded state are difficult to stabilize due to lack of intermediate information (short life and least stability). Therefore, stabilization of large proteins is the most difficult task as found from present research scenario. It has been found that three-dimensional structure of the protein has several clues to find the principle of stabilization strategies. In case of simple three-state folding equilibrium having single intermediate conformation (appearing at mildly denaturing conditions) existing between native and denatured state. Folding equilibrium is defined by following equation:

$$N \xrightarrow{\Delta G_{NI}} I \xrightarrow{\Delta G_{ID}} D$$

ΔG_{NI}: represents the stability of the native conformation relative to the intermediate
ΔG_{ID}: represents stability of denatured state relative to the intermediate
   Due to short life of intermediate, the 'relevant' conformational stability cannot be predicted. Thus, total conformational stability is given by ΔG_{NI} only. Further, due to lack of understanding of NI equilibria energetics, it is not clear whether the stabilization strategies found to stabilize two-state proteins will work well for proteins with equilibrium intermediates [119]. Therefore, folding energetics should be thoroughly studied before exploring stabilization strategies. Folding intermediates are sometimes difficult to see due to limitation of spectroscopic methods being used for monitoring unfolding/refolding. Other techniques based on the principle of calorimetry like differential scanning calorimeter have been found to be successful in various cases. However, superposition of analysis from various methods like fluorescence spectroscopy, circular dichroism, UV-VIS spectroscopy etc., would help

in getting better insight of understanding folding equilibria as well as φ-analysis of native state, folding intermediates and unfolded state of protein. These studies are also helpful in mutational studies of any protein by observing shift in T_ms of the *NI* and *ID* equilibria and changes in free energy caused due to mutation by using following equation:

$$\phi_I = \Delta\Delta G_{ID} / \Delta\Delta G_{ND}$$

When $\varphi_I = 1$, the broken interaction is as present in the intermediate as it is in the native state, and when φ_I is 0, then the interaction is absent in the intermediate. Other φ_I values, between 0 and 1, might in principle reflect a variety of intermediate conformations. Further, shape of the unfolding transition helps to predict whether the unfolding of the intermediate is cooperative or not. Thus, φ-analysis is helpful in studying stabilizing as well as destabilizing single mutations. It gives the information of interacting residues in the native state of the protein using double mutations. However, the analysis has limitation in various cases as follows [120]:

- The probed interaction formed in intermediate and native states which is protein stabilizing. However, removing or introduction of such interaction is not changing the 'relevant' stability.
- The interaction is protein stabilizing but it is not formed in the intermediate state. Removing such interaction decreases the 'relevant' stability, while engineering new such interaction increases the 'relevant' stability.
- The interaction is protein destabilizing (for example, an electrostatic repulsion between two neighboring charges) which is formed in the intermediate as well as in the native states. However, removing such interaction is not changing the 'relevant' stability.
- The interaction is protein destabilizing which is not formed in the intermediate. Removing such interaction increases the 'relevant' stability, while engineering new such interaction decreases it.

Therefore, researchers are still not able to anticipate type of interactions present in intermediate and native state in various cases. The solution to the problem is to study more and more complex proteins particularly their folding energetics to come with some definite conclusions. These studies will also be helpful to come out for the solution of protein misfolding which occur between folded and unfolded state.

Recently, hydrogen transfer method based on deuterium isotope has raised some hope to study complex proteins. Here exposed amide protons present on the surface of the protein are exchanged with deuterium present in the solvent while buried amide protons as well as those involved in intramolecular hydrogen bonding are protected. The structure and dynamics of a folded protein at equilibrium can be probed by monitoring the rate of exchange of amide protons with solvent deuterons. The hydrogen exchange is a two-step process involving both open and closed states [121]:

$$NH(closed) \underset{k_{cl}}{\overset{k_{op}}{\rightleftharpoons}} NH(open) \xrightarrow{k_{ch}} exchanged$$

Here, k_{op} and k_{cl} are the opening (unfolding) and closing (folding) rates of the protecting structure. The chemical exchange rate of freely available, unprotected amide hydrogens, k_{ch}; depends on variety of conditions (pH, temperature, neighboring amino acid side chains, and isotope effects). Under steady-state conditions, the exchange rate (k_{ex}) is determined by following equation:

$$k_{ex} = \frac{k_{op}k_{ch}}{k_{op} + k_{cl} + k_{ch}}$$

In most of the cases when $k_{op} \gg k_{cl}$, then above equation can be written as:

$$k_{ex} = \frac{k_{op}k_{ch}}{k_{cl} + k_{ch}}$$

This equation reduces to two limiting cases. Under *EX2* (bimolecular exchange) conditions where $k_{cl} \gg k_{ch}$ (low pH and temperature), the exchange rate becomes,

$$k_{ex}^{EX2} = \frac{k_{op}k_{ch}}{k_{cl}} = K_{op}k_{ch}, K_{op} = k_{op}/k_{cl}$$

The stabilization free energy of the protecting structure can then be calculated as

$$\Delta G_{HX} = -RT \ln K_{op} = -RT \ln(k_{ex}/k_{ch})$$

Under *EX1* (monomolecular exchange) conditions where $k_{cl} < k_{ch}$ (high pH, high temperature, low stability), the exchange rate limits at the opening rate of the protecting structure,

$$k_{ex}^{EX1} = k_{op}$$

These equations translate HX rates measured in *EX1* and *EX2* modes into information about the thermodynamics and kinetics of the protecting structure. The relationships that connect *EX1* and *EX2* behaviors are shown in Fig. 3.1. Exposed regions of the native state of the protein generally exchange with *EX2* kinetics while deeply buried regions in the structure of the protein exchange with *EX1* kinetics. Therefore, determination of particular proton exchange helps in the identification of relative location of that proton within the protein.

Following is one of the examples showing significance of hydrogen transfer in predicting protein folding pathways:

Amide Hydrogen Transfer in Cyt c [122] Three-dimensional structure of Cyt *c* has been shown in Fig. 3.2a. The structure can be distinguished based on the colors of loops and helix as yellow, blue, green and red. At low concentration of guanidium chloride (GdmCl), there is no exposure of buried protons. Thus, most of the

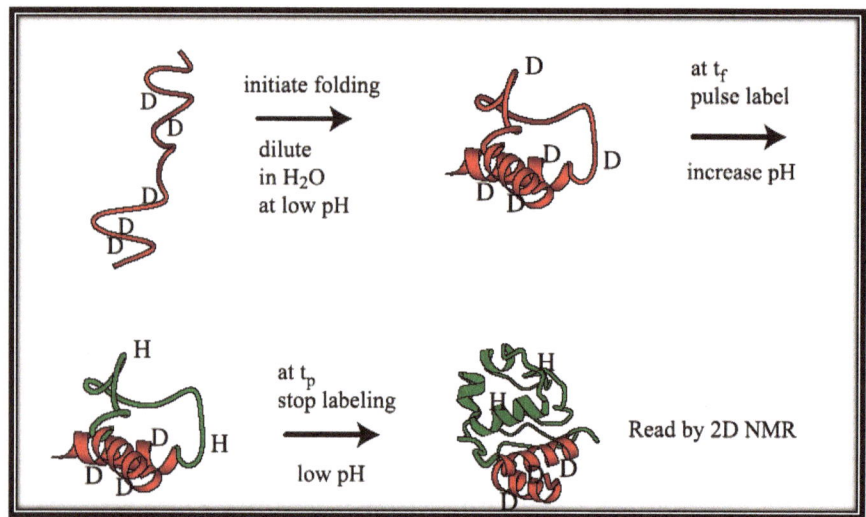

Fig. 3.1 HX pulse labeling experiment. Folding of polypeptide chain is proceeded using dilution method containing D_2O. When an intermediate accumulates in three-state kinetic folding, a brief H–D exchange labeling pulse can selectively label still unprotected sites. The sites that were protected and unprotected at the time of the pulse can be read using 2D-NMR analysis of the native protein (t_f: folding time, t_p: time of pulse). [Adapted from ref. 121]

amides exchange occurs through local fluctuations. This is shown by the lack of dependence of ΔG_{HX} on denaturant concentration ($m \sim 0$). With increase in the concentration of denaturant, buried protons gets exposed to the solvent, thus undergo hydrogen exchange *via* solvent containing D_2O. As shown in Fig. 3.2b, correlation of free energy ΔG_{HX} with increasing concentration of GdmCl. Slope of the linear

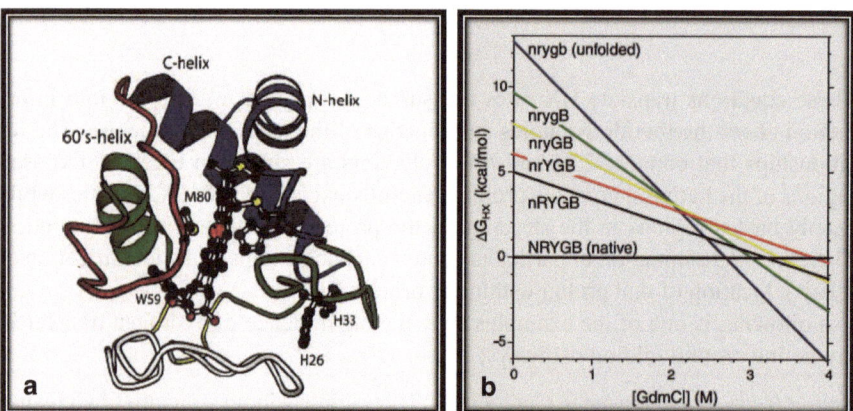

Fig. 3.2 Cooperative foldons in Cyt *c*, color coded in the structure (**a**) and in a crossover plot (**b**). Folding units (foldons) is defined by native state and pulse labeling HX experiments. During Cyt *c* folding, the foldons are sequentially put into one place at a time which subsequently progressed to native like partially unfolded forms (PUFs). The crossover plot shows the dependence on GdmCl concentration of the free energy levels of the increasingly structured PUFs. [Adapted from ref 121 and 122]

curves is directly proportional to surface exposure (m). The unfolding of five units (red loop, blue helix, yellow loop, green helix and green loop) can be found on the basis of hydrogen transfer method as:

$$NRYGB \rightleftharpoons nRYGB \rightleftharpoons nrYGB \rightleftharpoons nryGB \rightleftharpoons nrygB \rightleftharpoons nrygb$$

Alphabets in small refer to corresponding unfolded state. Sequential steps of unfolding follows: unfolding of Nested yellow loop (N) followed by Red loop (R), Yellow loop (Y), Green helix (G), and Blue helix (B). As these experiments are done at equilibrium native conditions, therefore pathway so revealed for unfolding is also applicable for refolding of Cyt c too. Each folding unit is termed as foldon representing a secondary structural unit or a group of one or two secondary elements. Here, the Cyt c protein is made of five foldons as its building blocks. There is sequential folding of each foldons followed by their inter-foldon interactions which are responsible for making native three-dimensional protein. The initial folding of the Blue unit provides a docking surface necessary to guide and stabilize the Green helix and loop which in turn are necessary to support the Yellow and Red loops. The principle as used here can be applied for various complex proteins, however may constrained for few cases due to limitation of techniques as discussed above.

The prediction of folding pathways as discussed here is useful for getting an interesting picture of protein structure. It seems that globular proteins are constructed of recognizably separate, individually cooperative building blocks called as foldons. The foldons maintain their separately cooperative nature within the native protein. Stepwise protein folding pathways arise as a consequence of this unit substructure and the way that the units interact in the native protein. These studies are helpful for understanding evolutionary studies of proteins as found in different species and would also help to some extent in prediction of three-dimensional structures from sequence of the protein.

Chapter 4
Involvement of Bioinformatics in Solving Protein Folding Problem

4.1 Introduction

There are lots of facts that have been revealed on protein folding problem since last decade. Most importantly, energy landscape theory and funnel concept have greatly evolved the field which describes protein folding is proceeded by progressive ensemble of partially folded structures leading to a completely folded native structure [123]. Protein folding is controlled by the shape of free energy landscape and the roughness on it arising due to various interactions which stabilize the folded state. Theoretical and experimental advances have found that designing of certain motifs can be done but there are various other motifs present in the protein which cannot be designed. This is due to fact that foldability of those motifs is independent of energetic frustrations. Further, prediction of order of native contact formation during folding is another challenge to be looked [124]. However, collective studies by theoretical folding studies, all-atom simulations and experimental evidences have suggested that the real proteins, particularly small, fast-folding two-state like proteins have sequences with a sufficiently reduced level of energetic frustration with transition state ensemble primarily determined by topological constraints. Topology is one of the dominant factors governing transition state structure and thus helpful in prediction of fold from the given sequence [125]. Recent observations have revealed that there is substantial correlation between the average sequence separation between contacting residues in the native structure and the folding rates for single domain proteins [126].

Protein folding is not characterized by single pathway but enormously large number of equivalent folding pathways reaching to the point of native state with single conformation of the backbone. Most challenging aspect toward theoretical biologists is to predict structure of transition state in protein folding. However, several models have been made based on Levinthal entropies, stabilities, and energetic roughness. These models have shown transition state ensemble lying about halfway through the unfolded and folded states [127]. The average amount of native formation in the transition ensemble using simulation is about 50 %. However, variations exist in the native structure derived from designed sequence *via* transition ensemble

© The Author(s) 2015
A. Dwevedi, *Protein Folding*, SpringerBriefs in Biochemistry and Molecular Biology,
DOI 10.1007/978-3-319-12592-3_4

due to variations in the amount of specific contacts found in the native protein. Real proteins display complex heterogeneity in contact formation due to which their exact structure is quite difficult to predict. Approach is being made to isolate sequences from database with minimal topological frustration to correlate them with minimalist models so that prediction of native contact can be done [128]. In addition to selecting sequences with lower energetic frustrations by theoreticians, evolution has also selected particular set of folding motifs which have reduced levels of topological frustration while discarding other structures difficult to fold [129]. The present chapter is based on understanding energy landscape and its significance toward prediction of three-dimensional structure of the real protein based on various topological parameters and native inter-residual contacts.

4.2 Understanding Energy Landscapes

The energy landscape is the mathematical correlation of energy of microscopic states to the macroscopic state. For example, in a system with *"n"* degrees of freedom having energy function given as:

$$F(x)=F(x_1, x_2,..., x_n),$$

where $x_1,..., x_n$ are variables specifying the microscopic state of the system.

In case of a protein, $x_1, x_2, ..., x_n$ are the dihedral angles of the polypeptide chain specifying a single conformation of the protein. Here, $F(x)$ is defined as the free energy of the given conformation of the protein, while entropy of each is calculated based on all possible solvent configurations. Therefore, energy landscapes can be defined as representation of the three-dimensional surfaces in case of proteins where vertical axis represents the free energy and the horizontal axis represents the conformational degrees of freedom of the polypeptide chain [130]. Proteins composed of amino acid sequence having a very rugged energy landscape due to ill defined stable conformation (Fig. 4.1). Further, most of the interactions in the proteins are formed between parts of the chain which are mutually supportive and cooperative leading to a low-energy structure, that is, "minimally frustrated". Due to minimal frustration, energy landscape of the protein takes the shape of funnel. A funnel-shaped energy landscape of protein infers that the free energy of the structure is minimal on achieving native state (most stable state) lying at bottom side of the funnel while top of the funnel is represented by the nonnative states with high conformational entropy, thus higher free energies [131].

A particular protein from any species would essentially have same structure and function. For example, sequences of adenylate kinase from pig and methanococcus (an archea bacterium) have homology of ~20%, however both are structurally similar. They have different thermodynamic and kinetic properties due to change in chain length and domain structure. This seems paradoxical, but it is true for number of proteins with their entries in protein data bank (PDB) (www.rcsb.org). Protein

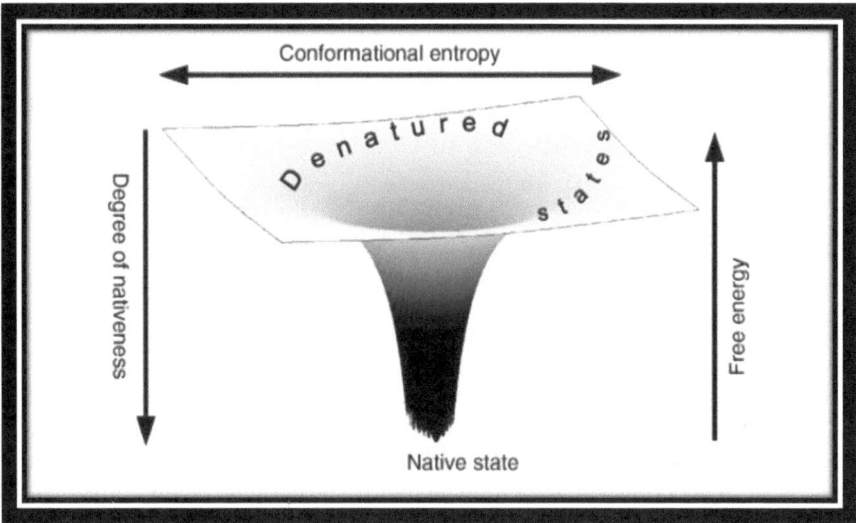

Fig. 4.1 Diagrammatic representation of a funnel-shaped energy landscape: The width of the funnel represents the conformational freedom of the polypeptide chain while vertical axis represents the free energy. The *bottom* of the funnel contains native state of the protein while unfolded states are present at the *top* of the funnel. (adapted from [144])

sequences look random and nonsymmetrical but their thorough analyses have revealed that they have many symmetrical structural elements. It is difficult to decipher structural relevance of given protein sequence unless it's three dimensional structure is known in PDB from other species. In order to find the solution to the problem, it is important to differentiate between thermodynamically foldable sequences and the subset of kinetically foldable sequences [132]. The understanding of energy landscapes, particularly the funneled nature of the landscape due to topological rearrangements which determine the uniqueness of the global fold present in protein sequences, from different species would be helpful. Several experiments have been done on analysis of energy landscapes models with the same protein with number of single point mutations. These experiments have found that single point mutation in the protein sequence lead to small perturbation to the energy landscape which could not change the basic shape of the funnel as there was no change in the global minimum. Thus, both mutated and wild type protein will fold into same three-dimensional structures. This structural robustness underlies the most commonly used method of predicting structures based on homology modeling [133].

The concept of energy landscapes also helps us to understand the rate and mechanism of protein folding in almost all small and large proteins. Protein folding involves slow steps, bottlenecks, and multiple kinetically distinguishable stages leading to formation of rugged funnel with small heaps and valleys. The exact layout of the small heaps and valleys on the energy landscape helps in understanding the details of the folding mechanism. For example, energy landscape view of hen egg white lysozyme folding has found that one region from the top of the funnel is smooth while another region has several bumps signifying one subpopulation of the molecules folds fast while another subpopulation folds quite slowly [134]. Further, the energy landscape view of protein folding is also helpful in understanding the mechanism of chaperone function to get a properly folded protein and its identification of misfolded protein [135].

4.3   Role of Energy Landscapes in Protein Folding

The folding landscape of proteins is rugged with several bumps due to presence of multiple conformations formed by various inappropriate contacts between residues of main and side chains. The formed structures with low-energy undergo conformational change which is directly proportional to $\sqrt{\Delta E^2}$. Folding landscape is more rugged at the top of the funnel than at the bottom which is characterized by stable native state of the protein with uniform topology. The physiologically native state of the protein has the lowest energy which is the purest state with slight changes in the side-chain orientations [136]. The parameters which are used to describe ensemble states in the funnel are: solvent-averaged energy (E), the fraction of native like contacts (Q), the percent of correct dihedral angles in the protein backbone (A), and the percent correct secondary structure. Further, folding landscape is stratified with each stratum arising due to various states with different energies.

The energy of a given misfolded state arises from the contributions of many random terms, so the probability distribution of energies at any stratum of the funnel is a Gaussian centered about the mean energy [137]:

$$P(Q,E) = \frac{1}{\sqrt{2\pi\Delta E^2(Q)}} \exp\{\frac{[E-\bar{E}(Q)]^2}{2\Delta E^2(Q)}\}$$

Unfolded protein with N residues having γ configurations per residue, then the total number of configurations (Ω_o) is directly proportional to γ^N. The total number of configurations decreases as the structures become more similar to the native state since there is only a single backbone conformation.

Two states of protein folding is characterized by U↔F, has free energy characterized by two minimum. One minimum is near to folded state and other to collapsed misfolded state with varying degrees of ordering. Slope of the energy landscape in two-state protein folding is given by following relation with assumption of probability of being folded is equal to probability of being in the misfolded state, i.e. $F_{native} = F(Q_{min}, T_F)$ and entropy of the folded protein is almost equal to its free energy. Here, T_f signifies folding temperature [138].

$$\delta E_S / T_F = S_o + \Delta E^2(Q_{min}) / 2k_B T_f^2$$

Further, $T_G(Q) = \sqrt{\dfrac{\Delta E^2(Q)}{2k_B S_o(Q)}}$

while T_G (glass-transition temperature), temperature where there are too few states available that system remains frozen with only few distinct states.

The ratio of T_F/T_G is used to distinguish fast and slow folding sequences and it is also important in predicting protein structures. This ratio is calculated using the set of states with the least structural similarity to the folded state:

$$\frac{T_F}{T_G} \approx \frac{\delta E_S}{\Delta E} \sqrt{\frac{2k_B}{S_o}}$$

Ratio of T_F/T_G is greater than 1 in case of fast folding while less than 1 in case of slow folding. It is independent of chain length but sensitive to interaction energy.

Energy landscapes clearly defines that there are various parameters besides number of contacts and their ordering are responsible for rate of protein folding. Folded protein has the core made of hydrophobic residues while surface is of hydrophilic residues. This ordering is brought by hydrophobic effect arising from protein folding in water (solvent). When the solvent is changed, it will effect the ordering of folding by effecting interaction energies of the hydrophobic groups. Due to this reason, under same physical conditions protein has random coil in one solvent while complete folded state in other solvent. Further temperature has strong control over hydrophobic forces directing ordering of folding [139, 140].

4.4 Computational Approaches to the Energy Landscapes of Protein Folding

The theory of energy landscape can be used to interpret the kinetics and thermodynamics of protein folding. It has been hoped that it will help in predicting the structure of a protein from its sequence and the design of novel proteins. The most important feature of energy landscape is its simplicity requiring only few parameters: stability gap between native and folded state of the protein, mean of the excited (misfolded) states and the roughness of the energy landscape. The model is primarily based on the fact that the folded conformation has the lowest energy with respect to all states with alternative conformations.

In case of structure prediction, the energy functions are based on [141]:

- Standard bonding and van der Waals interactions
- Knowledge-based potentials derived from known structures

The quantum based mechanical calculations are being tried to fit into spectroscopic data of small protein molecules to describe the motion of the atoms about or near their crystallographic coordinates. However, not much success has been achieved yet due to variation in time scale of atomistic simulations. Theoreticians are seeking simpler interaction which can encode sequence-structure relationships. According to the latest release of the SWISSPROT database, there are only 4000 proteins having their structure known from X-ray diffraction and nuclear magnetic resonance out of sequences of 40,000. Energy landscape theory has the clue of understanding the correlation of sequence-structure from these raw data. Scientists are trying to break the code of protein folding based on the information collected from sequence and structural databases along with such an organized theory of inference. However, it will also be requiring optimization of γ-parameters representing various configurations to achieve coherence of theoretical to experimental protein structures. Various optimization strategies have been adopted, *viz.* extracting structure-based potentials and correlating to experimentally observed frequencies of non-bonded amino acid residue pair contacts. The potential of mean force (W) between two residues in contact at a distance (r_c) is assumed to be related through a Boltzmann factor as given [142]:

$$\exp\{-W_{ij}(r_c)/RT\} = \frac{P_{ij}(r_c)}{P_{ij}(ref)}$$

Where, P_{ij} is a reference probability and T is the effective temperature.

Using the quasi-chemical approximation by neglecting chain connectivity, effective contact energy E_{ij} can be measured relative to either self-interaction as;

$$E_{ij} = W_{ij}(r_c) - [W_{ii}(r_c) + W_{jj}(r_c)]/2$$

Or interactions with some average residue or solvent molecule like "A"

$$E_{ij} = W_{ij}(r_c) + W_{AA}(r_c) - W_{iA}(r_c) - W_{jA}(r_c)$$

In the simplest application, the interaction sites are chosen to be at C_α atom or some atom in the side chain of each amino acid residue. The potentials of mean force using radial distribution functions $g_{ij}(r, l)$ for pairs of residues separated by a distance of l in the sequence is given by:

$$W_{ij}(r,l) = -RT \log g_{ij}(r,l)$$

These calculations are helpful in giving effective temperatures, with an assumption of minimum frustration along slope of funnel. However, experimentally mimicking similar conditions is too difficult in the present scenario. Although, various researchers across the world have achieved some success in generating native-like conformations with ~30% of the native contacts in place [143]. Various optimization strategies in addition to Monte Carlo or molecular dynamics are being used to predict protein structures. Threading algorithm is one of the practical procedure that tries to thread a new sequence onto all the known structures. It is based on the fact that there are certain superfolds which dominate protein structure database, *viz.* there are 200 different fold topologies in 4000 protein structures present in SWIS-SPROT. Threading tries to match structure with sequence in a fashion similar to the way sequences are matched onto each other in phylogenetic analysis. Many other similar schemes are also being developed, as recently a scheme based on potentials of mean force is given much priority. Here, low-energy misfolded structures are explicitly constructed to compute gap as well as ruggedness. The evaluation of sequence-structure alignment is based on contributions from pair contact. The total energy (E_T) is given by;

$$E_T = E_p + E_{ct} + E_{hb} + E_g + E_{exp}$$

Where,

Profile energy (E_p): measure of the propensity of an amino acid A_i to reside in a particular context of secondary structure SS_i and surface accessibility given by [144];

$$E_p = \sum_{i=1}^{N} \gamma^P (A_i, SS_i, SA_i)$$

Contact energy (E_{ct}): measures the pairwise interaction energies within two cut-off radii and monitors selected multibody interactions (mb) such as multiple cysteine bond formations [145]:

$$E_{ct} = \sum_{i=1}^{N-2} \sum_{j=i+2}^{N} \sum_{k=1}^{2} \gamma_k^{ct} (A_i, A_j) u(r_k^{ct} - r_{ij}) + mb$$

r_{ij} distance between residues A_i and A_j. In simple models u can be taken as unit step functions that include all short range interactions acting within 5 Å and all intermediate range interactions between 5 and 12 Å, respectively.

E_{hb} provides contributions for backbone hydrogen bond formation within α-helices and between β-sheets.

E_g gap energy which enforces only physically acceptable insertions, deletions, or bulges in the sequence-structure alignment.

E_{exp} used to aid in guiding an alignment to incorporate experimental data such as known contacts in an active site.

This algorithm has so far been used in threading due to easier steps for calculating low-energy minima. The self-consistent energy function produces low *rms* threading alignments for distant homologs. It is successful for cases like archeael adenylate kinase sequence, Fig. 4.2 [132].

The theory of protein folding based on energy landscapes provides a general framework for thinking about many facets of this complex problem. The formal theory of folding using energy landscapes is evolving with incorporation of several

Fig. 4.2 Predicted structure of adenylate kinase from *Methanococcus jannaschii*. The structure was obtained using mean-field alignment of the archaeal sequence with the scaffold of an uridylate kinase from yeast. The sequence–structure alignment is done using equation $E_T = E_p + E_{ct} + E_{hb} + E_g + E_{exp}$. Black portion shown in the structure represents the site for catalysis. (biosynthesis of ADP from ATP and AMP; adapted from [138])

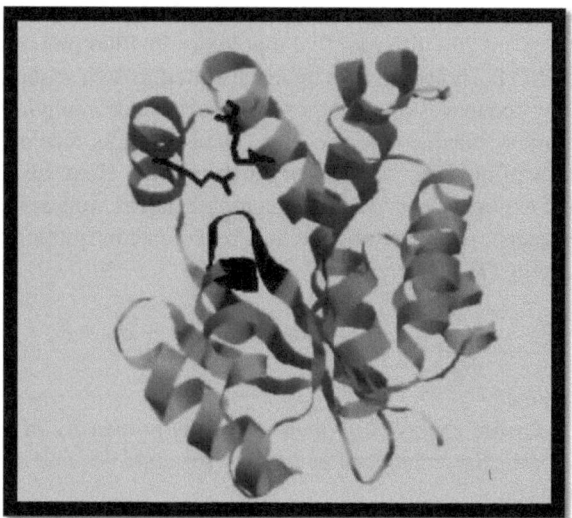

rectifications. Developments are being made that would help theoreticians to work at submillisecond scale to understand various stages of protein folding thoroughly. Protein engineering is complementing by undergoing several site-specific mutations to extract new information about the correlations of structures in partially folded protein with fully folded states. Protein structure prediction is reverse of protein engineering which is trying to extract information from the structure and incorporating to sequence design. Till date, success has been achieved in theoretical models and generated much scope for natural models. There is a need for development of reliable atomistic potentials to make it a real success in coming future.

4.5   Prediction of Protein Folding Pathways Using Various Servers

Computer simulations based on atomic physical force field have generated lots of scope in protein structure prediction and folding pathways. However, there is much to do in this respect to make it a real success of folding protein on computers. Major challenges raised toward theoreticians are [146]:

- Prediction of conformational changes (*viz.* induced fit)
- Understanding mechanisms requiring nonstatic native state of protein (*viz.* enzymatic catalysis, various stages of folding)
- Effect of chemical and physical factors on protein

Above mentioned problem are being thoroughly looked and sorted to few extents. The major success came through the development of fine grained distributed supercomputers having zest check point storage systems. Replica exchange molecular dynamics (REMD) is a most important tool used to sample conformations. Further, force field has particularly addressed backbone torsional energies which are helping in studying solvation as well as several interactions including ion pairing, salt bridging, etc. Recently, combination of conformational search technique based on zipping and assembly method (ZAM) of folding mechanism and AMBER96 force field have predicted native structures with residues ≤ 100. It involves following steps [147]:

a. Fragmentation of full protein chain into smaller fragments of ~8 mers and each fragment is simultaneously simulated using REMD.
b. Obtained metastable structures corresponding to each fragment are collected and zipped together followed by again simulation using REMD.
c. Stable inter- and intra-residual contacts are made using harmonic force field with limited degree of freedom.

This technique has been tested for folding of eight out of nine small proteins with 76–112 residues ($rms \leq 2.5$ Å) using a 70-processor cluster for over 6 months. The success has been achieved in four cases with roughly correct tertiary structures containing segments of ≥ 40 residues having an average RMSD of 5.9 Å. Further, secondary structures were also predicted with 73 % accuracy [148].

4.6   Future Prospects

Experimental as well as theoretical techniques have greatly helped in understanding the problem of protein folding. There are various facts which are revealed since last decades due to tremendous boost contributed by computational power. Experimental techniques like, hydrogen exchange and Φ-value methods can probe effect of mutations on folding rates, single-molecule method can explore heterogeneity of folding, energy landscapes and fast temperature-jump methods helpful in mimicking the folding pathway, etc. Theoretical and computational approaches including methods of bioinformatics, multiple-sequence alignments, force fields, models of energy landscapes, conformational sampling, structure-prediction Web servers, and fast protein loop sampling are being used to explore the physical mechanisms of folding while parallel and distributed grid-based computing are helpful for protein structure prediction [149, 150]. Thus, we have now reached the stage where we can simulate folding of few small proteins and even prediction of their three-dimensional structures can be done with more than 70 % accuracy. Further, we are capable of decoding the information store in amino acid sequence of protein, still we cannot predict structure directly from the sequence in absence of model systems. However, there are still many pillars in the problem of protein folding to be reached, *viz.* understanding folding behavior of complex multidomain proteins, protein–protein interactions, misfolding, aggregation, and amyloid formation.

However, there are various challenges in the field of protein folding which are still to be solved but we have come a long way where there were various insurmountable challenges which have been solved. Now, we have solution for Levinthal puzzle with the fact that the protein folds in microseconds rather in astronomical numbers as there is sequential folding of smaller units called foldons with subsequent inter-residual interactions leading to formation of completely folded native protein. Current knowledge of folding codes is helpful in designing new proteins as well as foldons using amino acid sequences. Therefore, being hopeful from the development of new experimental techniques and new theoretical approaches made in the field of protein folding it is possible to bring about great advances and solve the problem by coming decade.

References

1   Religa TL, Markson JS, Mayor U, Freund SMV, Fersht AR. Solution structure of a protein denatured state and folding intermediate. Nature. 2005;437:1053–6.
2   Frieden C, Hoeltzli SD, Ropson IJ. NMR and protein folding: equilibrium and stopped-flow studies. Prot Sci. 1993;2:2007–14.
3   Shiu Y-J, Jeng U-S, Huang Y-S, Lai Y-H, Lu H-F, Liang C-T, Hsu I-J, Su C-H, Su C, Chao I, Su A-C, Lin S-H. Global and local structural changes of Cytochrome c and lysozyme characterized by a multigroup unfolding process. Biophys J. 2008;94:4828–36.
4   Luo J, Maréchal J-D, Wärmländer S, Gräslund A, Perálvarez-Marín A. In Silico analysis of the apolipoprotein E and the amyloid β peptide interaction: misfolding induced by frustration of the salt bridge network. PLoS Comput Biol. 2010;6:e1000663.
5   Costandi M. Proteins behind mad-cow disease also help brain to develop: when not misfolded, prions lend a hand in forming neuronal connections. Nature. doi:10.1038/nature.2013.12428.
6   Muller PAJ, Vousden KH. p53 mutations in cancer. Nat Cell Biol. 2013;15:2–8.
7   Hopkins FG. Denaturation of proteins by urea and related substances. Nature. 1930;126:383–4.
8   Wu H. Studies on denaturation of proteins. XIII. A theory of denaturation. Chin J Physiol. 1931;5:321–44.
9   Rimington C. Protein structure and denaturation. Nature. 1931;127:440–1.
10 Speakman JB, Hirst MC. The pH stability region of insoluble proteins. Nature. 1931;127:665–6.
11 Fischer A. Heat denaturation of proteins as a chain reaction. Nature. 1936;137:576–7.
12 Young EG. Native state of proteins in egg-white. Nature. 1940;145:1021.
13 Bowen TJ. Physical studies on a soluble protein obtained by the degradation of elastin with urea. Biochem J. 1953;55:766–8.
14 Simpson RB, Kauzmann W. The kinetics of protein denaturation. I. The behavior of the optical rotation of ovalbumin in urea solutions. J Am Chem Soc. 1953;75:5139–52.
15 Levinthal C. Are there pathways for protein folding? Journal de Chimie Physique et de Physico-Chimie Biologique. 1968;65:44–5.
16 Wetlaufer DB. Nucleation, rapid folding, and globular intrachain regions in proteins. Proc Natl Acad Sci U S A. 1973;70:697–701.
17 Tanford C. Protein denaturation. Adv Prot Chem. 1968;23:121–282.
18 Reich S, Yonath J. Study of collagen denaturation kinetics by dynamic mechanical methods. Biopolymers. 1968;6:997–1000.
19 Goldsack DE. Volume fraction of the polypeptide chain in proteins and the concentration of protein denaturants. Biopolymers. 1968;6:164–5.
20 Milla ME, Brown BM, Sauer RT. P22 Arc repressor: enhanced expression of unstable mutants by addition of polar C-terminal sequences. Prot Sci. 1993;2:2198–205.

© The Author(s) 2015
A. Dwevedi, *Protein Folding*, SpringerBriefs in Biochemistry and Molecular Biology,
DOI 10.1007/978-3-319-12592-3

21   Brown BM, Milla ME, Smith TL, Sauer RT. Scanning mutagenesis of Arc repressor as a
      functional probe of operator recognition. Nature Struct Mol Biol. 1994;1:164–8.

22   Dudock BS, Katz G, Taylor EK, Holley RW. Primary structure of wheat germ phenylalanine
      transfer RNA. Proc Natl Acad Sci U S A. 1969;62:941–5.

23   Scheffler IE, Elson EL, Baldwin RL. Helix formation by d(TA) oligomers. II. Analysis of the
      helix-coil transitions of linear and circular oligomers. J Mol Biol. 1970;48:145–171.

24   Lewis PN, Go N, Go M, Kotelchuck D, Scheraga HA. Helix probability profiles of de-
      natured proteins and their correlation with native structures. Proc Natl Acad Sci U S A.
      1970;65:810–5.

25   Tsong TY, Baldwin RL, Elson EL. The sequential unfolding of ribonuclease A: detection of
      a fast initial phase in the kinetics of unfolding. Proc Natl Acad Sci U S A. 1971;68:2712–5.

26   Ikai A, Tanford C. Kinetic evidence for incorrectly folded intermediate states in the refolding
      of denatured proteins. Nature. 1971;230:100–2.

27   Brown JE, Klee WA. Helix-coil transition of the isolated amino terminus of ribonuclease.
      Biochemistry. 1971;10:470–6.

28   Garel JR, Baldwin RL. Both the fast and slow refolding reactions of ribonuclease A yield
      native enzyme. Proc Natl Acad Sci U S A. 1973;70:3347–51.

29   Brandts JF, Halvorson HR, Brennan M. Consideration of the possibility that the slow step in
      protein denaturation reactions is due to *cis-trans* isomerism of proline residues. Biochemis-
      try. 1975;14:4953–63.

30   Garel JR, Nall BT, Baldwin RL. Guanidine-unfolded state of ribonuclease A contains both
      fast- and slow-refolding species. Proc Natl Acad Sci U S A. 1976;73:1853–7.

31   Hagerman PJ, Baldwin RL. A quantitative treatment of the folding transition of ribonuclease
      A. Biochemistry. 1976;15:1462–73.

32   Blum AD, Smallcombe SH, Baldwin RL. Nuclear magnetic resonance evidence for a
      structural intermediate at an early stage in the refolding of ribonuclease A. J Mol Biol.
      1978;118:305–16.

33   Nall BT, Garel J-R, Baldwin RL. Test of the extended two-state model for the kinetic inter-
      mediates observed in the folding transition of ribonuclease A. J Mol Biol. 1978;118:317–30.

34   Schmid FX, Baldwin RL. Acid catalysis of the formation of the slow-folding species of ri-
      bonuclease A: evidence that the reaction is proline isomerization. Proc Natl Acad Sci U S A.
      1978;75:4764–8.

35   Lesk AM, Chothia C. Solvent accessibility, protein surfaces, and protein folding. Biophys J.
      1980;32:35–47.

36   Zettlmeiss G, Rudolph R, Jaenicke R. Limited proteolysis as a tool to study the kinetics of
      protein folding: conformational rearrangements in acid-dissociated lactic dehydrogenase as
      determined by pepsin digestion. Arch Biochem Biophys. 1983;224:161–8.

37   Roder H, Elöve GA, Englander SW. Structural characterization of folding intermediates in
      cytochrome *c* by H-exchange labelling and proton NMR. Nature. 1988;335:700–4.

38   Udgaonkar JB, Baldwin RL. NMR evidence for an early framework intermediate on the fold-
      ing pathway of ribonuclease A. Nature. 1988;335:694–9.

39   Hughson FM, Wright PE, Baldwin RL. Structural characterization of a partly folded apomyo-
      globin intermediate. Science. 1990;249:1544–8.

40   Hughson FM, Barrick D, Baldwin RL. Probing the stability of a partly folded apomyoglobin
      intermediate by site-directed mutagenesis. Biochemistry. 1991;30:4113–8.

41   Barrick D, Baldwin RL. Three-state analysis of sperm whale apomyoglobin folding. Bio-
      chemistry. 1993;32:3790–6.

42   Houry WA, Rothwarf DM, Scheraga HA. A very fast phase in the refolding of disulfide-
      intact ribonuclease A: implications for the refolding and unfolding pathway. Biochemistry.
      1994;33:2516–30.

43   Ybe JA, Kahn PC. Slow-folding kinetics of ribonuclease-A by volume change and circular
      dichroism: evidence for two independent reactions. Prot Sci. 1994;3:638–49.

44   Neira JL, Rico M. Folding studies on ribonuclease A, a model protein. Fold Des.
      1997;2:R1–R11.

45 Kiefhaber T, Labhardt AM, Baldwin RL. Direct NMR evidence for an intermediate preceding the rate-limiting step in the unfolding of ribonuclease A. Nature. 1995;375:513–5.

46 Hoeltzli S, Frieden C. Stopped-flow NMR spectroscopy: real-time unfolding studies of [6–19F] tryptophan-labeled *Escherichia coli* dihydrofolate reductase. Proc Natl Acad Sci U S A. 1995;92:9318–22.

47 Loh SN, Kay MS, Baldwin RL. Structure and stability of a second molten globule intermediate in the apomyoglobin folding pathway. Proc Natl Acad Sci U S A. 1995;92:5446–50.

48 Jamin M, Baldwin RL. Refolding and unfolding kinetics of the equilibrium folding intermediate of apomyoglobin. Nat Struct Mol Biol. 1996;3:613–8.

49 Luo Y, Kay MS, Baldwin RL. Cooperativity of folding of the apomyoglobin pH 4 folding intermediate studied by glycine and proline mutations. Nat Struct Mol Biol. 1997;4:925–30.

50 Lilie H, Haehnel W, Rudolph R, Baumann U. Folding of a synthetic parallel β-roll protein. FEBS Lett. 2000;470:173–7.

51 Basharov MA. Are synthetic proteins relevant to the problem of protein folding? Biofizika. 2002;47:989–95.

52 Oh K, Jeong KS, Moore JS. Folding driven synthesis of oligomers. Nature. 2001;414:889–93.

53 Kentsis A, Mezei M, Osman R. MC-PHS: a Monte Carlo implementation of the primary hydration shell for protein folding and design. Biophys J. 2003;84:805–15.

54 Cieplak M, Hoang TX. Universality classes in folding times of proteins. Biophys J. 2003;84:475–88.

55 Mayor U, Guydosh NR, Johnson CM, Grossmann JG, Sato S, Jas GS, Freund SM, Alonso DO, Daggett V, Fersht AR. The complete folding pathway of a protein from nanoseconds to microseconds. Nature. 2003;421:863–67.

56 Yun CH, Tang YH, Feng YM, An XM, Chang WR, Liang DC. 1.42 Å crystal structure of mini-IGF-1(2): an analysis of the disulfide isomerization property and receptor binding property of IGF-1 based on the three-dimensional structure. Biochem Biophys Res Commun. 2005;326:52–9.

57 Araç D, Dulubova I, Pei J, Huryeva I, Grishin NV, Rizo J. Three-dimensional structure of the rSly1 N-terminal domain reveals a conformational change induced by binding to syntaxin 5. J Mol Biol. 2005;346:589–601.

58 Grace CR, Durrer L, Koerber SC, Erchegyi J, Reubi JC, Rivier JE, Riek R. Somatostatin receptor 1 selective analogues: 4. Three-dimensional consensus structure by NMR. J Med Chem. 2005;48:523–33.

59 Bowie JU. Solving the membrane protein folding problem. Nature. 2005;438:581–9.

60 Sugase K, Dyson HJ, Wright PE. Mechanism of coupled folding and binding of an intrinsically disordered protein. Nature. 2007;447:1021–5.

61 Mok KH, Kuhn LT, Goez M, Day IJ, Lin JC, Andersen NH, Hore PJ. A pre-existing hydrophobic collapse in the unfolded state of an ultrafast folding protein. Nature. 2007;447:106–9.

62 Shank EA, Cecconi C, Dill JW, Marqusee S, Bustamante C. The folding cooperativity of a protein is controlled by its chain topology. Nature. 2010;465:637–40.

63 Hartl FU, Bracher A, Hayer-Hartl M. Molecular chaperones in protein folding and proteostasis. Nature. 2011;475:324–32.

64 Motlagh HN, Wrabl JO, Li J, Hilser VJ. The ensemble nature of allostery. Nature. 2014;508:331–9.

65 Lammert H, Wolynes PG, Onuchic JN. The role of atomic level steric effects and attractive forces in protein folding. Prot: Struct, Funct Bioinform. 2012;80:363–73.

66 Galzitskaya OV. Influence of conformational entropy on the protein folding rate. Entropy. 2010;12:961–82.

67 Ahmed SA, Ruvinov SB, Kayastha AM, Miles EW. Mechanism of mutual activation of the tryptophan synthase alpha and beta subunits. Analysis of the reaction specificity and substrate-induced inactivation of active site and tunnel mutants of the beta subunit. J Biol Chem. 1991;266:21548–57.

68 Batey S, Clarke J. Apparent cooperativity in the folding of multidomain proteins depends on the relative rates of folding of the constituent domains. Proc Natl Acad Sci U S A. 2006;103:18113–8.

69 Creighton TE. Protein folding: does diffusion determine the folding rate? Curr Biol. 1997;7:R380–R383.

70 Dwevedi A, Dubey VK, Jagannadham MV, Kayastha AM. Insights into pH-induced conformational transition of β-galactosidase from *Pisum sativum* leading to its multimerization. Appl Biochem Biotechnol. 2010;162:2294–312.

71 Niesen FH, Berglund H, Vedadi M. The use of differential scanning fluorimetry to detect ligand interactions that promotes protein stability. Nat Protoc. 2007;2:2212–21.

72 Walter S, Buchner J. Molecular chaperones-cellular machines for protein folding. Angew Chemie Int Ed. 2002;41:1098–113.

73 Guo L, Giasson BI, Glavis-Bloom A, Brewer MD, Shorter J, Gitler AD, Yang X. A cellular system that degrades misfolded proteins and protects against neurodegeneration. Mol cell. 2014;55:15–30.

74 Bessette PH, Aslund F, Beckwith J, Georgiou G. Efficient folding of proteins with multiple disulfide bonds in the *Escherichia coli* cytoplasm. Proc Natl Acad Sci U S A. 1999;96:13703–8.

75 Sridevi K, Lakshmikanth GS, Krishnamoorthy G, Udgaonkar JB. Increasing stability reduces conformational heterogeneity in a protein folding intermediate ensemble. J Mol Biol. 2004;337:699–711.

76 Mirny L, Shakhnovich E. Protein folding theory: from lattice to all-atom models. Annu Rev Biophys Biomol Struct. 2001;30:361–96.

77 Gromiha MM, Selvaraj S. Comparison between long-range interactions and contact order in determining the folding rate of two-state proteins: application of long-range order to folding rate prediction. J Mol Biol. 2001;310:27–32.

78 Istomin AY, Jacobs DJ, Livesay DR. On the role of structural class of a protein with two-state folding kinetics in determining correlations between its size, topology, and folding rate. Prot Sci. 2007;16:2564–9.

79 Ivankov DN, Finkelstein AV. Prediction of protein folding rates from the amino acid sequence-predicted secondary structure. Proc Natl Acad Sci U S A. 2004;101:8942–4.

80 Tripathi P, Hofmann H, Kayastha AM, Ulbrich-Hofmann R. Conformational stability and integrity of alpha-amylase from mung beans: evidence of kinetic intermediate in GdmCl-induced unfolding. Biophys Chem. 2008;137:95–9.

81 Ivankov DN, Garbuzynskiy SO, Alm E, Plaxco KW, Baker D, Finkelstein AV. Contact order revisited: influence of protein size on the folding rate. Prot Sci. 2003;12:2057–62.

82 Plaxco KW, Simons KT, Baker D. Contact order, transition state placement and the refolding rates of single domain proteins. J Mol Biol. 1998;277:985–94.

83 Gromiha MM. Importance of native-state topology for determining the folding rate of two-state proteins. J Chem Inf Comput Sci. 2003;43:1481–5.

84 Fersht AR. Transition-state structure as a unifying basis in protein-folding mechanisms: contact order, chain topology, stability, and the extended nucleus mechanism. Proc Natl Acad Sci U S A. 2007;97:1525–9.

85 Riddle DS, Grantcharova VP, Santiago JV, Alm E, Ruczinski I, Baker D. Experiment and theory highlight role of native state topology in SH3 folding. Nat Struct Mol Biol. 1999;6:1016–24.

86 Finkelstein AV, Bogatyreva NS, Garbuzynskiy SO. Restrictions to protein folding determined by the protein size. FEBS Lett. 2013;587:1884–90.

87 Finkelstein AV, Ivankov DN, Garbuzynskiy SO, Galzitskaya OV. Understanding the folding rates and folding nuclei of globular proteins. Curr Prot Pept Sci. 2007;8:521–36.

88 Privalov PL. Stability of proteins: small globular proteins. Adv Prot Chem. 1979;33:167–241.

89 Garbuzynskiy SO, Ivankov DN, Bogatyreva NS, Finkelstein AV. Golden triangle for folding rates of globular proteins. Proc Natl Acad Sci U S A. 2013;110:147–50.

90 Taylor WR, Chelliah V, Hollup SM, MacDonald JT, Jonassen I. Probing the "dark matter" of protein fold space. Structure. 2009;17:1244–52.

91   Murzin AG, Brenner SE, Hubbard T, Chothia C. SCOP: a structural classification of proteins database for the investigation of sequences and structures. J Mol Biol. 1995;247:536–40.

92   Orengo CA, Michie AD, Jones S, Jones DT, Swindells MB, Thornton JM. CATH-A hierarchic classification of protein domain structures. Structure. 1997;5:1093–108.

93   Gregersen N, Bross P, Vang S, Christensen JH. Protein misfolding and human disease. Annu Rev Genomics Hum Genet. 2006;7:103–24.

94   Makin OS, Serpell LC. X-ray diffraction studies of amyloid structure. Method Mol Biol. 2005;299:67–80.

95   Chamberlain AK, MacPhee CE, Zurdo J, Morozova-Roche LA, Hill HA, Dobson CM, Davis JJ. Ultrastructural organization of amyloid fibrils by atomic force microscopy. Biophys J. 2000;79:3282–93.

96   Stefani M, Rigacci S. Protein folding and aggregation into amyloid: the interference by natural phenolic compounds. Int J Mol Sci. 2013;14:12411–57.

97   Morel B, Varela L, Azuaga AI, Conejero-Lara F. Environmental conditions affect the kinetics of nucleation of amyloid fibrils and determine their morphology. Biophys J. 2010;99:3801–10.

98   Kowalska A. Amyloid precursor protein gene mutations responsible for early-onset autosomal dominant Alzheimer's disease. Folia Neuropathol. 2003;41:35–40.

99   Mandel-Gutfreund Y, Gregoret LM. On the significance of alternating patterns of polar and non-polar residues in beta-strands. J Mol Biol. 2002;323:453–61.

100   Borysik AJ, Morten IJ, Radford SE, Hewitt EW. Specific glycosaminoglycans promote unseeded amyloid formation from β_2-microglobulin under physiological conditions. Kidney Int. 2007;72:174–81.

101   Helmbrecht K, Zeise E, Rensing L. Chaperones in cell cycle regulation and mitogenic signal transduction: a review. Cell Prolif. 2000;33:341–65.

102   Bański P, Kodiha M, Stochaj U. Chaperones and multitasking proteins in the nucleolus: networking together for survival? Trend Biochem Sci. 2010;35:361–7.

103   Naylor DJ, Hartl FU. Contribution of molecular chaperones to protein folding in the cytoplasm of prokaryotic and eukaryotic cells. Biochem Soc Symp. 2001;68:45–68.

104   Young JC, Agashe VR, Siegers K, Hartl FU. Pathways of chaperone-mediated protein folding in the cytosol. Nat Rev Mol Cell Biol. 2004;5:781–91.

105   Green DE, Zande HD. Universal energy principle of biological systems and the unity of bioenergetics. Proc Natl Acad Sci U S A. 1981;78:5344–7.

106   Jacobs DJ, Dallakyan S. Elucidating protein thermodynamics from the three-dimensional structure of the native state using network rigidity. Biophys J. 2005;88:903–15.

107   Pace CN, Shirley BA, Mcnutt M, Gajiwala K. Forces contributing to the conformational stability of proteins. FASEB J. 1996;10:75–83.

108   Ooi T. Thermodynamics of protein folding: effects of hydration and electrostatic interactions. Adv Biophys. 1994;30:105–54.

109   Dwevedi A, Kayastha AM. Stabilization of beta-galactosidase (from peas) by immobilization onto amberlite MB-150 beads and its application in lactose hydrolysis. J Agric Food Chem. 2009;57:682–8.

110   Bakk A, Høye JS, Hansen A. Heat capacity of protein folding. Biophys J. 2001;81:710–4.

111   Fitter J. A measure of conformational entropy change during thermal protein unfolding using neutron spectroscopy. Biophys J. 2003;84:3924–30.

112   Dill KA. Dominant forces in protein folding. Biochemistry. 1990;29:7133–55.

113   Vertrees J, Wrabl JO, Hilser VJ. Energetic profiling of protein folds. Method Enzymol. 2009;455:299–327.

114   Baldwin RL. Energetics of protein folding. J Mol Biol. 2007;371:283–301.

115   Makhatadze GI, Privalov PL. On the entropy of protein folding. Prot Sci. 1996;5:507–10.

116   Kauzmann W. Factors in interpretation of protein denaturation. Adv Prot Chem. 1959;14:1–63.

117   Lee B, Richards FM. The interpretation of protein structures: estimation of static accessibility. J Mol Biol. 1971;55:379–400.

118 Liu Z, Chan HS. Solvation and desolvation effects in protein folding: native flexibility, kinetic cooperativity and enthalpic barriers under isostability conditions. Phys Biol. 2005;2:S75–S85.

119 Kumar R, Tripathi P, de Moraes FR, Caruso IP, Jagannadham MV. Identification of folding intermediates of streblin, the most stable serine protease: biophysical analysis. Appl Biochem Biotechnol. 2014;172:658–71.

120 Ozkan SB, Bahar I, Dill KA. Transition states and the meaning of ϕ-values in protein folding kinetics. Nat Struct Biol. 2001;8:765–9.

121 Krishna MMG, Hoang L, Lin Y, Englander SW. Hydrogen exchange methods to study protein folding. Methods. 2004;34:51–64.

122 Maity H, Maity M, Krishna MMG, Mayne L, Englander SW. Protein folding: the stepwise assembly of foldon units. Proc Natl Acad Sci U S A. 2005;102:4741–6.

123 Leeson DT, Gai F, Rodriguez HM, Gregoret LM, Dyer RB. Protein folding and unfolding on a complex energy landscape. Proc Natl Acad Sci U S A. 2000;97:2527–32.

124 Das P, Wilson CJ, Fossati G, Wittung-Stafshede P, Matthews KS, Clementi C. Characterization of the folding landscape of monomeric lactose repressor: Quantitative comparison of theory and experiment. Proc Natl Acad Sci U S A. 2005;102:14569–74.

125 Levy Y, Onuchic JN. Water and proteins: a love-hate relationship. Proc Natl Acad Sci U S A. 2004;101:3325–6.

126 Ouyang Z, Liang J. Predicting protein folding rates from geometric contact and amino acid sequence. Prot Sci. 2008;17:1256–63.

127 Shimada J, Shakhnovich EI. The ensemble folding kinetics of protein G from an all-atom Monte Carlo simulation. Proc Natl Acad Sci U S A. 2002;99:11175–80.

128 Nymeyer H, García AE, Onuchic JN. Folding funnels and frustration in off-lattice minimalist protein landscapes. Proc Natl Acad Sci U S A. 1998;95:5921–8.

129 Salunke DM, Gill J, Dwevedi A. Comparative structural proteomics of allergenic proteins from plant pollen. J Indian Inst Sci. 2014;94:119–26.

130 Gerstman BS, Chapagain PP. Self-organizing dynamics in protein folding. Prog Mol Biol Transl Sci. 2008;84:1–37.

131 Kaffe-Abramovich T, Unger R. A simple model for evolution of proteins towards the global minimum of free energy. Fold Des. 1998;3:389–99.

132 Shakhnovich EI, Gutin AM. Implications of thermodynamics of protein folding for evolution of primary sequences. Nature. 1990;346:773–5.

133 Xiang Z. Advances in homology protein structure modeling. Curr Prot Pept Sci. 2006;7:217–27.

134 Kulkarni SK, Ashcroft AE, Carey M, Masselos D, Robinson CV, Radford SE. A near-native state on the slow refolding pathway of hen lysozyme. Prot Sci. 1999;8:35–44.

135 Kim YE, Hipp MS, Bracher A, Hayer-Hartl M, Hartl FU. Molecular chaperone functions in protein folding and proteostasis. Annu Rev Biochem. 2013;82:323–55.

136 Ferreiro DU, Walczak AM, Komives EA, Wolynes PG. The energy landscapes of repeat-containing proteins: Topology, cooperativity, and the folding funnels of one-dimensional architectures. PLoS Comput Biol. 2008;4:e1000070.

137 Plotkin SS, Wolynes PG. Buffed energy landscapes: another solution to the kinetic paradoxes of protein folding. Proc Natl Acad Sci U S A. 2003;100:4417–22.

138 Onuchic JN. Theory of protein folding: the energy landscape perspective. Annu Rev Phys Chem. 1997;48:545–600.

139 Sundd M, Kundu S, Dubey VK, Jagannadham MV. Unfolding of ervatamin C in the presence of organic solvents: sequential transitions of the protein in the O-state. J Biochem Mol Biol. 2004;37:586–96.

140 Uversky VN, Li J, Fink AL. Evidence for a partially folded intermediate in α-synuclein fibril formation. J Biol Chem. 2001;276:10737–10744.

141 Lazaridis T, Karplus M. Effective energy functions for protein structure prediction. Curr Opin Struct Biol. 2000;10:139–45.

142   Kloczkowski A, Jernigan RL, Wu Z, Song G, Yang L, Kolinski A, Pokarowski P. Distance matrix-based approach to protein structure prediction. J Struct Funct Genomics. 2009;10:67–81.

143   Lezon TR, Bahar I. Using entropy maximization to understand the determinants of structural dynamics beyond native contact topology. PLoS Comput Biol. 2010;6:e1000816.

144   Sasai M, Wolynes PG. Unified theory of collapse, folding, and glass transitions in associative-memory Hamiltonian models of proteins. Phys Rev A. 1992;46:7979–97.

145   Finkelstein AV, Badretdinov AY, Gutin AM. Why do proteins architectures have boltzmann-like statistics? Prot: Struct Funct Bioinform. 1995;23:142–50.

146   Jones DT. Critically assessing the state-of-the-art in protein structure prediction. Pharmacogenomics J. 2001;1:126–34.

147   Shell MS, Ozkan SB, Voelz V, Wu GA, Dill KA. Blind test of physics-based prediction of protein structures. Biophys J. 2009;96:917–24.

148   Ahmadi Adl A Nowzari-Dalini A Xue B Uversky VN Qian X. Accurate prediction of protein structural classes using functional domains and predicted secondary structure sequences. J Biomol Struct Dyn. 2012;29:623–33.

149   Lutz B, Sinner C, Bozic S, Kondov I, Schug A. Native structure-based modeling and simulation of biomolecular systems per mouse click. BMC Bioinform. 2014;15:292–304.

150   Wassenaar TA, van Dijk M, Loureiro-Ferreira N, van der Schot G, de Vries SJ, Schmitz C, van der Zwan J, Rolf Boelens R, Giachetti A, Ferella L, Rosato A, Bertini I, Herrmann T, Jonker RAH, Bagaria A, Jaravine V, Güntert P, Schwalbe H, Vranken WF, Doreleijers JF, Vriend G, Vuister GW, Franke D, Kikhney A, Svergun DI, Fogh RH, Ionides J, Laue ED, Spronk C, Jurkša S, Verlato M, Badoer S, Mazzucato SD, Frizziero E, Bonvin AM. WeNMR: structural biology on the grid. J Grid Comput. 2010;10:743–67.